Cables and Wiring, Second Edition

John Cadick and
AVO International

Delmar Publishers

an International Thomson Publishing company 

Albany • Bonn • Boston • Cincinnati • Detroit • London • Madrid
Melbourne • Mexico City • New York • Pacific Grove • Paris • San Francisco
Singapore • Tokyo • Toronto • Washington

Cover Design: Charles Cummings Advertising Art, Inc.

Delmar Staff:
Publisher: Alar Elken
Acquisitions Editor: Mark Huth
Project Editor: Megeen Mulholland
Production Manager: Mary Ellen Black
Art and Design Coordinator: Cheri Plasse
Editorial Assistant: Dawn Daugherty

Printed in the United States of America

For more information contact:

Delmar Publishers
3 Columbia Circle, Box 15015
Albany, New York 12212-5015

International Thomson Publishing Europe
Berkshire House
168–173 High Holborn
London, WC1V7AA
United Kingdom

Nelson ITP, Australia
102 Dodds Street
South Melbourne,
Victoria, 3205 Australia

Nelson Canada
1120 Birchmont Road
Scarborough, Ontario
M1K 5G4, Canada

International Thomson Publishing France
Tour Maine-Montparnasse
33 Avenue du Maine
75755 Paris Cedex 15, France

International Thomson Editores
Seneca 53
Colonia Polanco
11560 Mexico D. F. Mexico

International Thomson Publishing GmbH
Königswinterer Straße 418
53227 Bonn
Germany

International Thomson Publishing Asia
60 Albert Street
#15–01 Albert Complex
Singapore 189969

International Thomson Publishing Japan
Hirakawa-cho Kyowa Building, 3F
2-2-1 Hirakawa-cho, Chiyoda-ku,
Tokyo 102, Japan

ITE Spain/Paraninfo
Calle Magallanes, 25
28015-Madrid, Espana

5 6 7 8 9 10 XXX 03 02 01

Library of Congress Cataloging-in-Publication Data
Cadick, John.
Cables and wiring / John Cadick and AVO International.—2nd ed.
p. cm.
Includes index.
ISBN 0-7668-0270-1 (pbk. : alk. paper)
1. Electric wiring. 2. Electric cables. 3. Electric wire. I. AVO International Training Institute. II. Title.
TK3201.C33 1999
621.319'24—dc21

98-40832
CIP

Table of Contents

PART TWO

Includes specific cable application information, such as cable sizes, current capacities, permitted uses, and applications.

Preface to the Second Edition

The second edition of *Cables and Wiring* was written as a reference book for journeymen and apprentice electricians. This book provides fingertip access to power and communications cable and wire construction, splicing, termination, testing, and applications.

Written in two parts, the second edition of *Cables and Wiring* begins with chapters that illustrate, in detail, acceptable methods for installing, splicing, and terminating power and communications cables.

The second part of *Cables and Wiring* is a complete reference to over 30 types of Underwriters Laboratories® and *National Electrical Code® (NEC®)* listed cable and wire. Information provided includes available sizes, ampacities, temperature ranges, allowable installations, receiving and handling, and termination methods.

Wire is the most common type of electrical equipment. When electricity is compared to water, wires are the pipes that pass the "electrical fluid" from where it is to where we want it to be. The electrical conductor compares to the opening in the pipe, and the insulation is similar to the wall of the pipe.

While the principles of wire are simple, diverse applications and service requirements have led to a wide variety of wire types, sizes, shapes, and structures. In addition to the different sizes for various loads, industry has also developed wire for wet environments, wire for corrosive environments, wire for hot environments, and wire for cold environments. Each of the different types of wire and cable has different handling, installation, and testing requirements.

All this complexity led to confusion and misunderstanding. The problem was aggravated by the fact that there were no good, concise references available for the practicing electrician, until now!

Intended for the practicing journeyman and apprentice electrician, *Cables and Wiring* provides information that is unavailable in any other single location. Starting with a simple yet comprehensive explanation of the fundamentals, the book advances to chapters detailing cable materials and construction, installation, testing and fault location, and a detailed synopsis of over thirty *NEC®* wire and cable types. Sections are included on application, sizes, temperature ranges, handling and care, installation, splicing and terminations, and testing for each of the major *NEC®* cable types.

When the electrician has questions about a particular type of wire and cable, he or she need only look as far as the information chapters at the end of this handbook. *Cables and Wiring* is destined to become the electrician's guide for wiring, its types, uses, applications, terminations, and testing.

Foreword

AVO International is a world leader in training services and electrical testing products. The Training Institute is a leading provider of hands-on technical skills training for electrical power system testing and maintenance. AVO Training and The Cadick Corporation have worked together on many projects since 1986.

A registered professional engineer, John Cadick has specialized for three decades in electrical engineering, training, and management. In 1986 he created Cadick Professional Services (forerunner to the present-day Cadick Corporation), a consulting firm in Garland, Texas. His firm specializes in electrical engineering and training, working extensively in the areas of power system design and engineering studies, condition based maintenance programs, and electrical safety. Prior to the creation of Cadick Corporation, John held a number of technical and managerial positions with electric utilities, electrical testing firms, and consulting firms. Mr. Cadick is a widely published author of numerous articles and technical papers. He is the author of the *Electrical Safety Handbook* as well as *Cables and Wiring*. His expertise in electrical engineering as well as electrical maintenance and testing coupled with his extensive experience in the electrical power industry makes Mr. Cadick a highly respected and sought after consultant in the industry.

Acknowledgments for the Second Edition

AVO International® and John Cadick, PE, wish to thank the following companies and individuals for their invaluable assistance in supplying technical information and art for this book.

A.J. Pearson	National Joint Apprenticeship Training Committee
Jim Boyd	National Joint Apprenticeship Training Committee
Jack Pullizzi	AT&T Bell Laboratories
Sam Brown	Rome Cable Corporation
L. James Milne	Pyrotenax USA, Incorporated
S. Graeme Thomson	AVO Biddle® Instruments
James Marshall	3M Electrical Products Division
Christopher B. Jonnes	Polywater Corporation
John Fee	Polywater Corporation
Alan Mark Franks	AVO International®
Michael D. McHorse, PE	The Okonite Company
Priscilla West	Raychem Corporation
John Cavenaugh	General Cable Company
Allan P. Marconi	Manhattan Electric Cable Corporation
Richard Aure	Canoga-Perkins Corporation
Christine Arnold	AMP Incorporated

I would like to extend special thanks to Donald J. Sterling, Jr. for his excellent book *Premises Cabling.* Donald's book was used extensively (and shamelessly) for material on communications cables. *Premises Cabling* is undoubtedly the best single reference source available for the subject of communications cabling and is highly recommended to the reader.

John Cadick

Dedication

To my wife, Sheryl, whose support, help, and confidence made this possible.
This book is dedicated to all the hardworking people at AVO International.

Delmar Publishers Is Your Electrical Book Source!

Whether you're a beginning student or a master electrician, Delmar Publishers has the right book for you. Our complete selection of proven best-sellers and all-new titles is designed to bring you the most up-to-date, technically-accurate information available.

NATIONAL ELECTRICAL CODE

National Electrical Code® 1999/NFPA

Revised every three years, the *National Electric Code®* is the basis of all U.S. electrical codes.

Order # 0-8776-5432-8

Loose-leaf version in binder

Order # 0-8776-5433-6

National Electrical Code® Handbook 1999/NFPA

This essential resource pulls together all the extra facts, figures, and explanations you need to interpret the 1999 *NEC®*. It includes the entire text of the Code, plus expert commentary, real-world examples, diagrams, and illustrations that clarify requirements.

Order # 0-8776-5437-9

Illustrated Changes in the 1999 National Electrical Code®/O'Riley

This book provides an abundantly-illustrated and easy-to-understand analysis of the changes made to the 1999 *NEC®*.

Order # 0-7668-0763-0

Understanding the National Electrical Code®, 3E/Holt

This book gives users at every level the ability to understand what the *NEC®* requires, and simplifies this sometimes intimidating and confusing code.

Order # 0-7668-0350-3

Illustrated Guide to the National Electrical Code®/Miller

Highly-detailed illustrations offer insight into Code requirements and are further enhanced through clearly-written, concise blocks of text that can be read very quickly and understood with ease. Organized by classes of occupancy.

Order # 0-7668-0529-8

Interpreting the National Electrical Code®, 5E/Surbrook

This updated resource provides a process for understanding and applying the *National Electrical Code®* to electrical contracting, plan development, and review.

Order # 0-7668-0187-X

Electrical Grounding, 5E/O'Riley

Electrical Grounding is a highly illustrated, systematic approach for understanding grounding principles and their application to the 1999 *NEC®*.

Order # 0-7668-0486-0

ELECTRICAL WIRING

Electrical Raceways and Other Wiring Methods, 3E/Loyd

The most authoritative resource on metallic and nonmetallic raceways, this book provides users with a concise, easy-to-understand guide to the specific design criteria and wiring methods and materials required by the 1999 *NEC®*.

Order # 0-7668-0266-3

Electrical Wiring Residential, 13E/Mullin

Now in full color! Users can learn all aspects of residential wiring and how to apply them to the wiring of a typical house from this, the most widely-used residential wiring book in the country.

Softcover Order # 0-8273-8607-9

Hardcover Order # 0-8273-8610-9

House Wiring with the NEC®/Mullin

The focus of this new book is the applications of the *NEC®* to house wiring.

Order # 0-8273-8350-9

Electrical Wiring Commercial, 10E/Mullin and Smith

Users can learn commercial wiring in accordance with the *NEC®* from this comprehensive guide to applying the newly revised 1999 *NEC®*.

Order # 0-7668-0179-9

Electrical Wiring Industrial, 10E/Smith and Herman

This practical resource has users work their way through an entire industrial building—wiring the branch circuits, feeders, service entrances, and many of the electrical appliances and subsystems found in commercial buildings.

Order #0-7668-0193-4

Cables and Wiring, 2E/AVO

This concise, easy-to-use book is your single-source guide to electrical cables—it's a "must-have" reference for journeyman electricians, contractors, inspectors, and designers.

Order # 0-7668-0270-1

ELECTRICAL MACHINES AND CONTROLS

Industrial Motor Control, 4E/Herman and Alerich

This newly revised and expanded book, now in full color, provides easy-to-follow instructions and essential information for controlling industrial motors. Also available are a new lab manual and an interactive CD-ROM.

Order # 0-8273-8640-0

Electric Motor Control, 6E/Alerich and Herman

Fully updated in this new sixth edition, this book has been a long-standing leader in the area of electric motor controls.

Order # 0-8273-8456-4

Introduction to Programmable Logic Controllers/Dunning

This book offers an introduction to Programmable Logic Controllers.

Order # 0-8273-7866-1

Technician's Guide to Programmable Controllers, 3E/Cox

Uses a plain, easy-to-understand approach and covers the basics of programmable controllers.

Order # 0-8273-6238-2

Programmable Controller Circuits/Bertrand

This book is a project manual designed to provide practical laboratory experience for one studying industrial controls.

Order # 0-8273-7066-0

Electronic Variable Speed Drives/Brumbach

Aimed squarely at maintenance and troubleshooting, *Electronic Variable Speed Drives* is the only book devoted exclusively to this topic.

Order # 0-8273-6937-9

Electrical Controls for Machines, 5E/Rexford

State-of-the-art process and machine control devices, circuits, and systems for all types of industries are explained in detail in this comprehensive resource.

Order # 0-8273-7644-8

Electrical Transformers and Rotating Machines/Herman

This new book is an excellent resource for electrical students and professionals in the electrical trade.

Order # 0-7668-0579-4

Delmar's Standard Guide to Transformers/Herman

Delmar's Standard Guide to Transformers was developed from the best-seller *Standard Textbook of Electricity* with expanded transformer coverage not found in any other book.

Order # 0-8273-7209-4

DATA AND VOICE COMMUNICATION CABLING AND FIBER OPTICS

Complete Guide to Fiber Optic Cable System Installation/Pearson

This book offers comprehensive, unbiased, state-of-the-art information and procedures for installing fiber optic cable systems.

Order # 0-8273-7318-X

Fiber Optics Technician's Manual/Hayes

Here's an indispensable tool for all technicians and electricians who need to learn about optimal fiber optic design and installation as well as the latest troubleshooting tips and techniques.

Order # 0-8273-7426-7

A Guide for Telecommunications Cable Splicing/Highhouse

A "how-to" guide for splicing all types of telecommunications cables.

Order # 0-8273-8066-6

Premises Cabling/Sterling

This reference is ideal for electricians, electrical contractors, and inspectors needing specific information on the principles of structured wiring systems.

Order # 0-8273-7244-2

ELECTRICAL THEORY

Delmar's Standard Textbook of Electricity, 2E/Herman

This exciting full-color book is the most comprehensive book on DC/AC circuits and machines for those learning the electrical trades.

Order # 0-8273-8550-1

Industrial Electricity, 6E/Nadon, Gelmine, and Brumbach

This revised, illustrated book offers broad coverage of the basics of electrical theory and industrial applications. It is perfect for those who wish to be industrial maintenance technicians.

Order # 0-7668-0101-2

EXAM PREPARATION

Journeyman Electrician's Exam Preparation, 2E/Holt

This comprehensive exam prep guide includes all the topics on journeyman electrician competency exams.

Order # 0-7668-0375-9

Master Electrician's Exam Preparation, 2E/Holt

This comprehensive exam prep guide includes all the topics on master electrician's competency exams.

Order # 0-7668-0376-7

REFERENCE

ELECTRICAL REFERENCE SERIES

This series of technical reference books is written by experts and designed to provide the electrician, electrical contractor, industrial maintenance technician, and other electrical workers a source of reference information about virtually all the electrical topics they encounter.

Electrician's Technical Reference—Motor Controls/Carpenter

Electrician's Technical Reference—Motor Controls is a source of comprehensive information on understanding the controls that start, stop, and regulate the speed of motors.

Order # 0-8273-8514-5

Electrician's Technical Reference—Motors/Carpenter

Electrician's Technical Reference—Motors builds an understanding of the operation, theory, and applications of motors.

Order # 0-8273-8513-7

Electrician's Technical Reference—Theory and Calculations/Herman

Electrician's Technical Reference—Theory and Calculations provides detailed examples of problem-solving for different kinds of DC and AC circuits.

Order # 0-8273-7885-8

Electrician's Technical Reference—Transformers/Herman

Electrician's Technical Reference—Transformers focuses on the theoretical and practical aspects of single-phase and 3-phase transformers and transformer connections.

Order # 0-8273-8496-3

Electrician's Technical Reference—Hazardous Locations/Loyd

Electrician's Technical Reference—Hazardous Locations covers electrical wiring methods and basic electrical design considerations for hazardous locations.

Order # 0-8273-8380-0

Electrician's Technical Reference—Wiring Methods/Loyd

Electrician's Technical Reference—Wiring Methods covers electrical wiring methods and basic electrical design considerations for all locations, and shows how to provide efficient, safe, and economical applications of various types of available wiring methods.

Order # 0-8273-8379-7

Electrician's Technical Reference—Industrial Electronics/Herman

Electrician's Technical Reference—Industrial Electronics covers components most used in heavy industry, such as silicon control rectifiers, triacs, and more. It also includes examples of common rectifiers and phase-shifting circuits.

Order # 0-7668-0347-3

RELATED TITLES

Common Sense Conduit Bending and Cable Tray Techniques/Simpson

Now geared especially for students, this manual remains the only complete treatment of the topic in the electrical field.

Order # 0-8273-7110-1

Practical Problems in Mathematics for Electricians, 5E/Herman

This book details the mathematics principles needed by electricians.

Order # 0-8273-6708-2

Electrical Estimating/Holt

This book provides a comprehensive look at how to estimate electrical wiring for residential and commercial buildings with extensive discussion of manual versus computer-assisted estimating.

Order # 0-8273-8100-X

Electrical Studies for Trades/Herman

Based on *Delmar's Standard Textbook of Electricity*, this new book provides nonelectrical trades students with the basic information they need to understand electrical systems.

Order # 0-8273-7845-9

PART ONE

Chapter 1

General Concepts

KEY POINTS

- What are conductors, insulators, and semiconductors?
- How does current flow through conductors, insulators, and semiconductors?
- What are the key thermal, electrical, and physical characteristics of electrical insulators?
- How does light travel through a material? How is it contained inside an optical fiber?

INTRODUCTION

Overview

Power or information can be transmitted in one of two ways. So-called "wireless" communications uses electromagnetic fields to send information through space from one antenna to another. Such transmission takes advantage of the fact that electromagnetic waves can pass through space with relatively little attenuation. The light from the stars we see at night travels via electromagnetic waves.

A generally more reliable method is to use some physical conductor to contain the information and pass it from one location to another. Examples of this type of transmission include:

- Power transmission using electrical conductors in lines or power cables.
- Information transfer via electrical conductors such as telegraph or computer networks.
- Information transfer using light waves along a fiber optic cable.

Note that there are only two transmission methods—electrical current and light waves.

Electrical Principles

All materials can be categorized based on their abilities to conduct electricity.

Conductors are those materials that offer little opposition to the passage of electric current. Materials such as copper or aluminum have large numbers of "free" electrons in their structure; when a voltage is applied, the electrons move easily. The electrical resistance of such a material is a few hundredths of an ohm per one thousand feet.

In contrast, insulators have relatively few free electrons in their atomic structure. Materials such as rubber and plastic allow very little current flow when a voltage is impressed across them and may have resistances of several hundred million ohms per inch.

Semiconductors fall between conductors and insulators in electrical resistance. Over a limited temperature range, semiconductors behave more like conductors than insulators. Although semiconductors share some of the properties of conductors, they do not conduct electricity as well. Semiconductor materials include silicon and germanium. These materials, when treated chemically, exhibit behavior that allows them to be used to make devices such as transistors and integrated circuits.

All of these materials are used in the construction of various types of wire and cable. See Chapter 2 for a more detailed description of wire and cable construction.

Light Principles

Light may also be considered an electromagnetic wave, differing from the electrical currents discussed previously only in frequency. Like electrical current, light passes through some objects easily, passes through others with more difficulty, and is blocked by yet others.

When light passes from one material to another, for example when it passes from water and enters air, it bends or refracts. Depending on the properties of the two materials, the light loses little energy while passing between the materials.

As light passes through a material, it is attenuated; that is, it is reduced in magnitude. Light behaves much like electricity as it passes through materials and, like electricity, it can be used to carry information from one end of a light conductor (called a fiber optic cable) to another.

Electrical Conductors versus Light Conductors

The amount of energy a fiber optic cable can carry is relatively small. Because of this, fiber optic cable is not used to transfer power. Because of its very high frequency, however, fiber optic cable can transmit enormous amounts of information. This capability makes fiber optics generally better than electrical conductors for information systems. Both fiber optic cable and electrical conductors are used to transmit information. Only electrical conductors are used to transmit power.

TERMS

The following terms and phrases are used in this chapter. (Many of these definitions are paraphrased from *ANSI/IEEE Std 100–1988: Standard Dictionary of Electrical and Electronic Terms.*)

Attenuation: The reduction of magnitude as a signal travels through a conductor or a fiber optic cable.

Cladding: The covering of a fiber optic cable. The primary purpose of the cladding is to serve as a refractive layer to keep the light beam within the fiber.

Conductor: A substance (usually metallic) that allows a current of electricity to pass continuously along it. This word may also be used to identify a wire or cable that is used in a power system to conduct electricity.

Core: The center portion of a fiber optic cable. The core carries the light beam from one end of the cable to the other.

Corona: The luminous discharge caused by the ionization of air around a high voltage conductor.

Critical Angle: The angle at which there is total internal reflection (refraction) of light energy. Above the critical angle, some or all of the light energy will pass through the interface of the two materials.

Dielectric Strength: The voltage gradient at which insulation breakdown occurs and current flows.

Equivalent Circuit: An arrangement of circuit elements that have characteristics electrically equivalent to, but may be physically different from, those of a different circuit or device.

Insulator: A material that impedes or opposes the passage of electric current flow.

Optical Spectrum: The frequency (or wavelength) band that contains optical light frequencies. Generally, the optical spectrum includes all frequencies from infrared to ultraviolet. The visible spectrum falls approximately in the middle of the optical spectrum.

Refraction: The action of bending light as it moves from one material to another.

Refraction Index: The ratio of the speed of light in a vacuum (or air) to the speed of light in a material.

Semiconductor: An electronic conductor with resistivity in the range between conductors and insulators.

Surface Current: Current that flows over the surface of an insulator.

Tracking: Degradation of insulation by the formation of external carbonized paths. The carbon tracks cause surface leakage current and can lead to catastrophic failure of the insulation.

Treeing: Degradation of insulation by the formation of internal carbonized paths. When examined in a cross-sectional view, the carbonized paths resemble the branches of a tree.

Volumetric Current: The current that flows through insulation.

CURRENT FLOW

Introduction

Ohm's law is the formula that defines the basic concept of current flow. Ohm's law states that the flow of current in an electric circuit is proportional to the applied voltage and inversely proportional to the resistance. In mathematical terms, Ohm's law may be shown as:

$$I = \frac{E}{R} \quad (1)$$

Where:

I = the current flow in amperes
E = the applied voltage in volts
R = the circuit resistance in Ohms

Although Ohm's law applies in virtually every situation, conditions may be different depending on the type of material and whether it is an insulator or a conductor. When used in electrical power applications, semiconducting material behaves very much like conducting material.

Conductors and Semiconductors

Equivalent Circuit. An equivalent circuit is a diagram that is used to describe the electrical behavior of a system. The equivalent circuit for a conductor is very simple and is shown in Figure 1–1.

As the current *(I)* flows through the wire, a voltage drop *(E)* is created according to Ohm's law. The resistance *(R)* will be very low for a good conductor.

Thermal Behavior. The resistance of all materials varies with temperature. A conductor has many free electrons. As the temperature goes up, the electrons vibrate more actively. Because of this, the electrons literally "get in each other's way" and cause the number of collisions between them to increase. The increase in collisions translates into more opposition to the current flow and, therefore, higher resistance.

Insulators

Equivalent Circuit. As you can see from Figure 1–2, the equivalent circuit for insulators is somewhat different from that for a conductor. Figure 1–2 shows the path the electrons will take as they pass *through* an insulator. The total current is composed of two components: the capacitive current (I_C) and the resistive current (I_R).

Figure 1–1 A conductor and its equivalent circuit.
(Courtesy of Cadick Corporation)

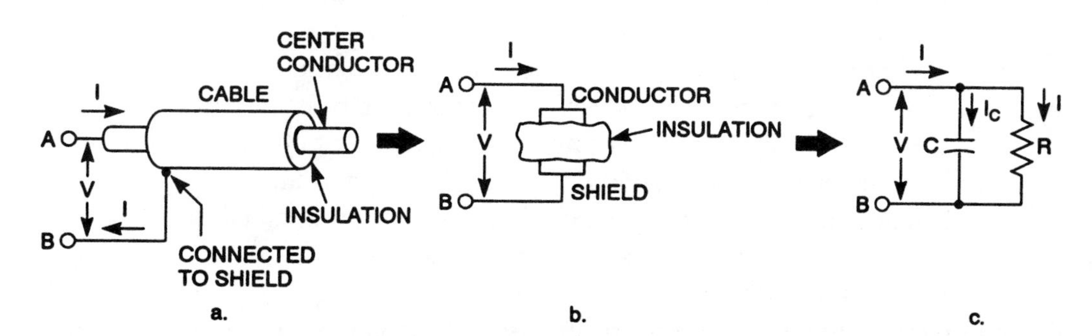

Figure 1–2 Insulation and its equivalent circuit.
(Courtesy of Cadick Corporation)

Capacitive Current. A capacitor is created when two conductors are separated by a layer of insulation, the exact situation that exists in a wire or cable. Virtually all electrical power circuits exhibit some degree of capacitance; wire and cable, however, are particularly capacitive.

When a capacitor has an alternating voltage impressed on it, the resulting current leads the applied voltage by 90°. The capacitive version of Ohm's law is:

$$I_C = \frac{E}{X_C} \angle 90° \tag{2}$$

Where:

I_C = the capacitive current flow
E = the applied voltage
X_C = the capacitive reactance

X_c is the capacitive equivalent of resistance. It is a function of geometric characteristics, such as the cross-sectional area of the conductive plates and separation of the conductive plates.

The capacitive current is approximately 100 times greater than the resistive current in good insulation.

Resistive Current. The resistive portion of the current flow, through and over the surface of an insulator, follows Ohm's law as explained in Equation (1). It is the result of electrons moving through and over the insulator in the same way that electrons move through a conductor.

The current that flows through the insulation is called the volumetric current. The volumetric current is caused by the motion of electrons through the insulation material. The part that flows over the surface is called the surface current. Surface current is caused by the motion of electrons through the contamination on the surface of the insulator.

Because there are far fewer free electrons in the insulator than in the conductor, the resistance of an insulator is much higher than the resistance of a conductor. The resistive current is approximately 1/100 of the capacitive current in good insulation. Resistive current in insulation is also called the leakage current.

The leakage current that flows through the insulation is called the volumetric leakage. The surface current is called the surface leakage.

Total Current. The total current in an insulator is equal to the vector sum of the capacitive current and the resistive current. Because the capacitive current in good insulation is so much greater than the resistive current, the total current is predominantly capacitive. Figure 1–3 is a vector diagram of the summation of the two currents.

As insulation ages, its resistive current will tend to increase; however, because the capacitive current is a function of geometry, its magnitude will remain essentially the same throughout the life of the insulation. Because good insulation has capacitive current flow that is typically 100 times its resistive current, testing the relationship between the resistive current and the capacitive current is one way that insulators can be evaluated for continued serviceability. A change in the ratio of capacitive current to resistive current indicates a problem with the insulation.

Thermal Behavior. As insulation is heated, its electrons become more energetic and therefore easier to displace. This results in a lowered resistance. This is one of the key electrical differences between conductors and insulators.

- When a conductor is heated, its resistance rises. The resistance of a conductor is proportional to temperature.
- When an insulator is heated, its resistance lowers. The resistance of an insulator is inversely proportional to temperature.

LIGHT FLOW

Light can be considered an electromagnetic wave of very high frequency. The frequencies we call light (the optical spectrum) fall into a frequency range of approximately 1×10^{14} Hz to slightly over 1×10^{15} Hz. (Note that 10^{14} is a 1 followed by

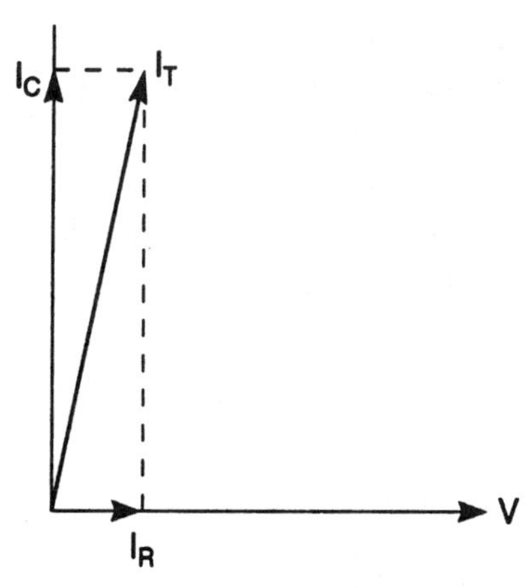

Figure 1–3 Vector sum of insulation capacitive current and resistive current. *(Courtesy of Cadick Corporation)*

14 zeros.) The visible spectrum (those frequencies we can see) falls into a frequency range of approximately 4.29×10^{14} Hz to 7.5×10^{14} Hz. In contrast, the FM radio spectrum covers from 88×10^6 Hz to 108×10^6 Hz.

Frequency versus Bandwidth

Because of the very large numbers involved, light frequencies are not normally expressed using hertz. Rather, the term "wavelength" is preferred. Wavelength and frequency are related by Equation (3):

$$\lambda = \frac{v}{c} \times 10^9 \qquad \textbf{(3)}$$

Where:

λ = wavelength in nanometers (10^{-9} meters)
c = the speed of light (3×10^8 meters per second)
v = frequency in hertz

Using nanometers is particularly useful, because it relates directly to the way in which fiber optic cable functions. This topic is covered in greater detail in Chapter 2.

Refraction

Light does not travel at the same speed in all media. Rather, it tends to be fastest in free space and air and slower in materials like glass, water, and plastic. Because of this fact, light tends to bend when it passes from one medium to another. This bending action is called refraction. Perhaps you have tried to pick up a coin from the bottom of a swimming pool. When you dove in, you may have been surprised to learn that the coin was not where it had appeared to be. This illusion is caused by the refraction of the light as it passes from the water in the pool to the air.

Refractive Index. The amount of bending that occurs is relatively predictable by using a unit called the "refractive index," as defined in Equation (4):

$$n = \frac{c_{vac}}{c_{mat}} \quad \textit{or, more commonly,} \quad n = \frac{c_{air}}{c_{mat}} \qquad \textbf{(4)}$$

Where:

n = refractive index
c_{vac} = the speed of light in a vacuum

c_{mat} = the speed of light in the material
c_{air} = the speed of light in air

Note that c_{vac} and c_{air} are very close in value, so that the refractive index of air is 1.00029 at room temperature.

When light strikes a material with a lower refractive index, some of the light will be refracted in the material and some will be reflected back into the original material. Below a certain angle, called the critical angle, virtually all the light will be reflected back into the medium with the higher refractive index.

Fiber Cable. The basic construction of fiber cable is shown in Figure 1–4.

The core of the cable has a higher refractive index than the cladding. Therefore, when a light beam strikes the cladding at an angle less than the critical angle, it is refracted back into the core to continue its travel. Typical sizes and designs are shown in Chapter 2. Most fiber optic systems used today are designed for wavelengths of from 800 nanometers to 1,600 nanometers. These frequencies are slightly below the visible spectrum.

Attenuation

Attenuation in fiber optic cable is very similar to the voltage drop created by the resistance of an electrical wire. As light travels through the cable, its magnitude is reduced, or attenuated. On very long runs of fiber, the signal may have to be boosted periodically using amplifiers. The unit of attenuation is the decibel, usually abbreviated dB. A three dB attenuation means that approximately one-half of the power has been lost.

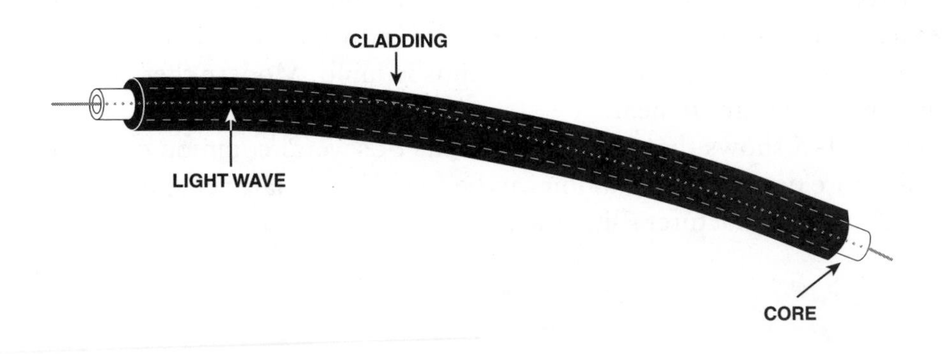

Figure 1–4 Fiber optic cable construction.
(Courtesy of Cadick Corporation)

Designing a fiber optic cable to minimize attenuation is a significant effort on the part of design engineers. When designed properly, fiber has a significantly lower attenuation rate than do electrical cables at the same frequency.

ENVIRONMENTAL EFFECTS

Overview

Various environmental conditions can have an adverse effect on the performance of electrical insulation. As a result, some types of cable that are resistant to these conditions have been developed. For example, silicone and Teflon® insulation have very high temperature characteristics, rubber insulation and lead sheathed cable are very resistant to moisture, and cross-linked polyethylene cable is resistant to chemical exposure. Specific information on the various cable types may be found in the reference chapters for the specific insulation and cable systems.

Heat

When insulation is heated, its physical characteristics change depending on the type of material that is being used. Some materials will become more flexible, while others may fail completely. If the heat is excessive, some insulation will allow the conductor to move, deforming the entire cable and degrading the insulating characteristics.

Some thermoplastic cable is also subject to a condition called treeing. Thermal and mechanical stress can cause degraded channels to form in the insulation. These channels resemble trees and allow current to flow, carbonizing them and further aggravating the problem. Eventually, the cable will fail completely.

In recognition of these problems, the *National Electrical Code® (NEC®)*[1] places maximum operating temperature limits on insulation. Manufacturers design cable insulation with these temperature limits in mind. Modern thermosetting insulation is more resistant to heat.

Figure 1–5 shows the temperature limits of several common cable insulation systems. More detailed information can be found in the later chapters of the book, the *NEC®*, or manufacturer's literature.

1 *National Electrical Code®* and *NEC®* are registered trademarks of the National Fire Protection Association.

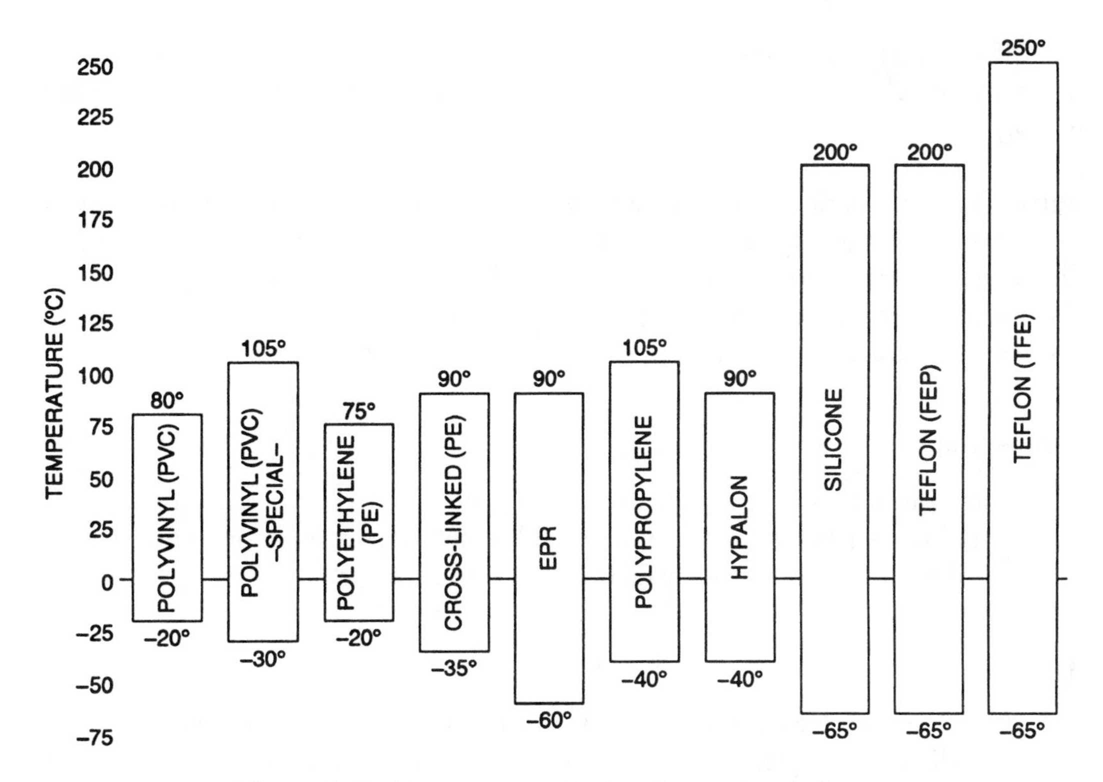

Figure 1–5 Temperature limits of some insulations.
(Courtesy of Manhattan Electrical Cable Corporation, Rye, New York)

Moisture

Surface. Surface moisture combined with contamination (dirt) will cause an increase in the surface leakage. These increased currents can cause two types of problems.

A phenomenon known as tracking can occur. This happens when the surface leakage currents heat the surface of the insulation, causing it to turn into carbon. As the surface tracks, the currents can increase, thereby aggravating the problem. Eventually, the cable can fail completely. This is a similar phenomenon to treeing, which was discussed in the last section. The principal difference is that treeing occurs inside the insulation, while tracking occurs on the surface. Tracking can occur on virtually all insulation materials, while treeing occurs primarily in the thermoplastic types of insulation.

Tracking and the resultant breakdown may be accompanied by corona. Corona is the luminous discharge caused by the ionization of the air around an energized conductor with a very high voltage gradient. Corona is often accompanied by the strong smell of ozone.

If moisture and contamination create a low enough resistance, the cable may fail immediately. Such a failure can be dramatic and extremely dangerous at any voltage.

Internal. All insulation materials can absorb moisture to some degree. As the moisture content increases, the insulation resistance and dielectric strength tend to decrease. Excessive moisture levels can cause the insulation to fail completely.

Moisture content is a special problem with the liquid insulation used in transformers and circuit breakers, but it can cause failure in solid insulation as well.

Chemical

Some insulation is adversely affected by chemicals. Acids, salts, alkalis, oils, and other hydrocarbons can cause some cable insulation to crack or distort. Chemical reaction can also lower the insulation dielectric strength.

Dirt

Dirt, dust, and other impurities become problems to insulation when they are mixed with moisture. The resulting mix of dirt and moisture creates conductive paths. This is a special problem if the contamination is near the end of the cable, close to the termination.

Chapter 2

Cable Materials and Construction

KEY POINTS

- How are low- and medium-voltage cables constructed?
- What types of materials are used for electrical cable conductors? What are their characteristics?
- What types of materials are used for electrical cable insulation? What are their characteristics?
- What types of materials are used for cable jackets? What are their characteristics?
- How are electrical and optical communications cables constructed?
- What are the operating characteristics of electrical and optical communications cables?
- What are the advantages of fiber optical cable?

INTRODUCTION

Overview

The construction of a given type of wire or cable depends on what the wire or cable is intended to do. Fiber optic cables, for example, have no copper. They are designed to carry light waves, not electrical ones. Coaxial communications cable, medium voltage power cable, and shielded twisted-pair cable all have external shields, but the ultimate applications of these very different types of cable determine the uses of those shields.

Keep in mind that the purpose of all types of cable is to transmit power and/or information from one end of the cable to the other. In power applications, the cable must keep the energy safely inside the cable to protect personnel and minimize interference from the power frequencies in use.

Similarly, communications cables must transmit information from one end of the cable to the other. In this case, however, the communications cable must minimize the leakage of information in both directions. That is, the information in the cable must be kept inside the cable, while external noise must be shielded so it does not interfere with the information.

Electrical Power Cable

Power cable construction and materials are determined by a variety of factors, including:

- Application voltage
- Load current
- Ambient temperature
- Ambient moisture
- Environmental chemicals
- Installation configuration

The history of power cable manufacturing and application is a history of changing requirements and innovation. In the early days, the industry began with simple paper and cloth insulation, but now the industry uses synthetic materials. Along the way, paper, oil-impregnated paper, varnished cambric, and other such insulations were (and in some cases are) extremely popular.

Copper has long been, and still remains, the "king" of the conductors. Relative low cost, comparative chemical stability, and low resistance all combine to make copper the ongoing favorite. When metallurgically combined with other materials, such as silver, copper becomes even better suited to the job.

Of course, the materials used are only part of the picture. The materials must be formed into a usable, rugged, reliable package that will carry the electric energy to the desired location and keep the electricity out of undesired locations.

Electrical Communications Cable

The types of communications cables popular today use many of the same construction materials found in power cable. Copper is the main conductor used with insulations made of synthetic plastics. The two most common forms of communications cables are coaxial and twisted pair.

Coaxial cable was introduced during the second world war and is used primarily for high frequency and/or high power communications transmission requirements. The coaxial cable relies primarily on its shield to prevent the leakage of information and/or the intrusion of external noise.

Twisted-pair cables, both shielded and unshielded, have become increasingly popular as the main means of interconnecting (networking) computer systems. The twisting of the pair adds to the security of the circuit by minimizing the amount of radiation the cable emits and the amount of radiation the cable receives. Shielding twisted conductors provide even more immunity to signal loss or outside interference.

Fiber Optic Cable

Fiber optic conductors do not use copper. Rather, they are made of glass, plastic, or some other similar material that passes light easily. Because of the very high frequencies employed in fiber optic channels, these channels are inherently impervious to electromagnetic interference. Therefore, fiber communications are used in many situations in which high reliability is required.

TERMS

The following terms and phrases are used in this chapter:

Ampacity: The amount of current a conductor will carry without exceeding an allowable temperature rise. The ampacity is specified at a given set of ambient conditions.

Anneal: To heat and slowly cool to toughen and reduce brittleness.

Buffer: The outermost layer of an optical fiber. The buffer provides the fiber mechanical protection.

Characteristic Impedance: The impedance a communications cable exhibits to a high frequency signal. Characteristic impedance is determined primarily by the size and shape of the cable and the insulation's dielectric constant; see definition of dielectric constant.

Circular Mil: The area of a circle that is 1 mil (0.001 inch) in diameter. Cable sizes are usually given in kcmils; see definition of kcmil (MCM).

Cladding: The layer that covers the core of an optical fiber. The cladding reflects the light waves and keeps them inside the core.

Coated Wire: Copper wire that has been coated with a tinning compound. Coated wire was originally developed to make copper more impervious to acids, which were used to prepare the wire for rubber coating.

Core: The centermost part of an optical fiber. The core contains and carries the light waves from the transmission source to the receiver.

dB: Abbreviation for deciBel. A deciBel is a comparison between two power levels. A three (3) deciBel loss (–3 dB) equals a ½ power loss.

Dielectric Constant: A characteristic of insulation that is associated with the force charge carriers exert on one another when the insulator is impressed with an electric field.

Ductile: Capable of being drawn out, as into wire, or being hammered thin.

Extrude: To make a material and give it certain properties by forcing it through a die in a liquid or semi-liquid state.

Hard-drawn: Conductors that are drawn through dies after they have cooled from the heating process.

Hygroscopic: Readily absorbing or retaining moisture.

kcmil (MCM): One thousand circular mils; the current preferred designation is kcmil and not MCM.

Low Voltage:

a. IEEE definition: Less than 1,000 volts AC.
b. *NEC*® definition: Less than 2,000 volts AC. This definition applies to cable and cable construction requirements. This handbook will use the *NEC*® definition unless otherwise stated.

Malleable: Capable of being shaped or formed, as by hammering or pressure.

Medium Voltage:

a. IEEE definition: 1,001 volts to 100,000 volts AC.
b. *NEC*® definition: 2,001 volts to 35,000 volts AC. This definition applies to cable and cable construction requirements. This handbook will use the *NEC*® definition unless otherwise stated.

Mode: The path light follows as it travels down a fiber. Fibers can be designed for as few as one mode or for many modes. Generally, the fewer the modes, the more data the fiber can carry (i.e., more bandwidth).

No-Oxide: A chemical treatment put on a metal to prevent oxidation.

Ozone: A form of oxygen with three atoms per molecule (O_3) rather than two (O_2). Ozone is much more chemically active than O_2 and destroys many types of insulation. Ozone is formed by high energy electrical discharges, such as arcing or corona.

Plenum: A compartment or chamber to which one or more air ducts are connected. The plenum forms the part of the air distribution system for a structure. Temperature extremes are quite high in a plenum; consequently, when they are used to route cabling, the cable must be high temperature or "plenum" rated.

RG/xU: The abbreviation used in MIL-17-C for coaxial cable. "RG" stands for Radio General and "U" stands for Utility (e.g., RG/6U or RG/17U).

Soft-Drawn: Conductors that are annealed by treating with heat and cooled to remove internal stress.

Thermoplastic: An insulating material that is soft when heated and hard when cooled. Examples include polyvinyl chloride (PVC) and polyethylene (PE).

Thermosetting: An insulating material similar to thermoplastic but with better thermal characteristics. Thermosetting material does not soften as easily as thermoplastic and is not as prone to permanent damage when heated. Cross-linked polyethylene (XLP) is a thermosetting insulation.

Uncoated Wire: Wire that has not been coated with a tinning compound.

Vulcanization: A process that increases the strength, elasticity, and resistance of a material. It usually involves using sulfur or other additives, along with heat and pressure.

ELECTRICAL CONDUCTORS

Conductor Types

Conductors are grouped into two basic types: solid and stranded. A solid conductor is made of a single strand of hard-drawn or soft-drawn wire. Hard-drawn conductors are wires that are mechanically drawn through dies after they have cooled from the heating process. Hard-drawn wire has a high tensile strength and a low amount of elongation under stress. This type of conductor is used primarily in outdoor applications, such as transmission lines. Hard-drawn conductors are very stiff and hard to bend and work.

Soft-drawn conductors are wires that are annealed by treating with heat and cooled to remove internal stress by making the metal less brittle. The soft-drawing process actually softens the metal (copper or aluminum) until it can be easily worked. Soft-drawn wire has low tensile strength and the greatest elongation under stress. Soft-drawn conductor wires bend easier than hard-drawn types and are used for indoor applications.

Solid conductors are primarily used in lighting, service, appliances, and some types of control systems. They are also used in some grounding systems. Solid conductors are difficult to work with in sizes larger than No. 8 AWG. Solid conductors in No. 12 AWG and below work well on screw terminals.

A stranded conductor is composed of single, solid wires twisted together. These single strands of wire may be twisted in a common bundle or individual groups.

Stranded conductors are used more frequently than solid conductors because they are very flexible. It is easier to use them to make connections, such as terminating at device connections and terminal boards. Extra flexibility can be obtained by using many strands of fine wire. Stranded conductors are easy to work with in all sizes, and they ensure good connection.

Stranded conductors also offer slightly less resistance than an equivalently sized solid conductor.

Conductor Materials

Introduction. Table 2–1 lists the materials and characteristics of the more commonly used conductors. The primary materials in use today are copper and aluminum. These two materials are used because they are reasonable in cost and are good electrical conductors.

Table 2–1 Conductor materials.

GOOD CONDUCTORS

Silver
- best conductor
- used for contacts
- plated onto some conductors

Copper
- good conductor
- most widely used

Aluminum
- not as good as copper
- strong, light, and cheap
- used for transmission conductors
- used for power distribution

Gold
- almost as good as copper
- extremely resistant to corrosion

- too expensive for general usage
- used primarily for contacts and terminals on printed circuit boards

FAIR CONDUCTORS

Brass, zinc, iron, and nickel
Tungsten (used for filaments)
Nichrome (used for heating elements)
Mercury (used for thermostats and switches)

Copper: Copper is the most important and commonly used of all conductor materials. Copper is used because it possesses several desirable characteristics.

- Copper is highly conductive (both thermally and electrically). It is second only to silver and gold in conductivity.
- Copper is plentiful and relatively inexpensive when compared to silver and gold.
- Copper is both ductile and malleable. These properties make it ideal for use as a conductor.
- Because it is resistant to both corrosion and fatigue, copper may be used in a variety of industrial and commercial environments.

Aluminum. Aluminum is the second most popular material used in the fabrication of electrical conductors. It is cheaper and lighter than copper and has almost as good thermal and electrical conductivity. Unlike copper, aluminum does not possess high tensile strength. In transmission lines, aluminum is reinforced with steel to give it the tensile strength required. Aluminum possesses several other characteristics.

- Aluminum expands and contracts on copper terminals, causing high resistance and resultant heat. High resistance builds up because the copper and aluminum have different thermal coefficients; therefore, the terminals loosen when aluminum is used with copper.
- Aluminum corrodes when connected to copper conductors because of galvanic chemical action caused by the reaction of the two dissimilar metals. No-oxide chemical compounds must be used to prevent this reaction.

- Aluminum is used in power systems, depending upon the design and code requirements.
- Aluminum is used in transmission, switchyard, building, and service cable.
- Aluminum does not conduct as well as copper and, therefore, must have a slightly larger cross-sectional area for the same ampacity. In spite of this, aluminum conductors are generally lighter and less expensive than copper conductors of the same ampacity.

Combination Materials. In wiring applications where high strength is required, a number of other conductors are available for use. These include coated copper, cadmium copper, chrome copper, and zirconium copper and are discussed as follows.

- When copper wire is coated with a tinning compound, it is easier to solder and is more impervious to corrosion. Coated wire was originally developed to protect copper from the acid compound that was used to prepare it for covering with rubber insulation.
- Nickel combined with copper strengthens the conductor, makes soldering easier, and reduces corrosive effects. Copper-coated nickel conductors may be used in power, control, and lighting systems.
- Copper covered steel is also available as either a fully annealed or hard-drawn material, the latter being preferred where high conductor strength is required. The conductance of these wires is generally specified as either 30 or 40 percent of an equivalently sized copper conductor at low or medium frequencies.
- Zirconium coated copper provides excellent high temperature performance.

Silver and Gold. Where maximum conductance (minimum resistance) is required, silver is widely used. Silver is the best conductor of *all* naturally occurring materials. Because it is quite expensive, it is used in limited applications, such as silver-plated contacts.

Although silver is an excellent conductor, it is quite soft and subject to corrosion and oxidation. Because of this, gold is often used in applications where silver is not acceptable. Because it is so expensive, gold is used in smaller electronic applications. The following summarizes the characteristics of these metals when used in electrical conducting service.

- Silver and especially gold are relatively expensive.
- Both materials are used primarily for contacts and other such limited applications.
- Gold is used when the corrosive tendencies of silver will cause problems.

Conductor Sizing

Conductors are sized with a wire gauge from high numbers down to size 0 (read "aught"). Size 0 is referred to as 1/0, pronounced "one-aught." Large numbers indicate small sizes of wire. For instance, a size 36 conductor is very, very fine, about the size of a strand of hair. A size 12 conductor is as large as the lead of a standard pencil.

Most conductors smaller than 1/0 can be measured with an American wire gauge (AWG), as illustrated in Figure 2–1. Conductors larger than 1/0 are sized with multiple zeros that increase with size; therefore, a 3/0 conductor is larger than a 2/0 conductor. Conductors are measured with a micrometer caliper. The wire size is usually identified on the insulation or jacket covering.

The cross-sectional area of conductors smaller than 250 kcmil is shown in Table 2–2.

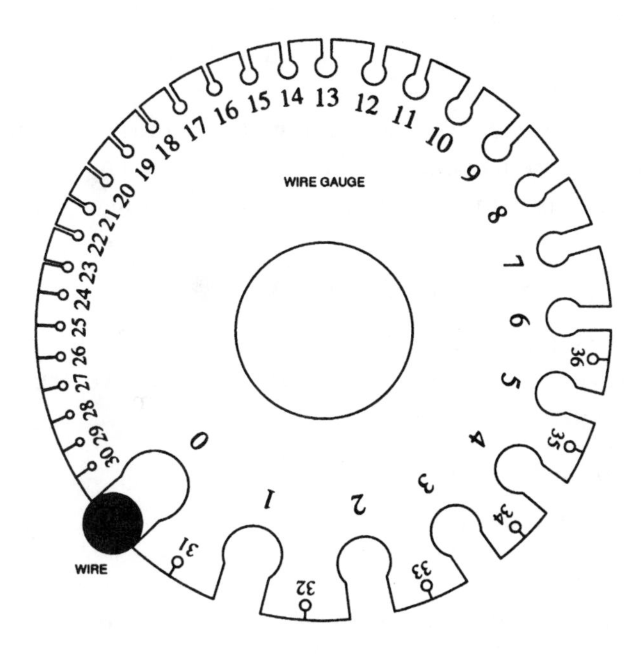

Figure 2–1 American Wire Gauge (AWG).

Table 2–2 Cross-sectional area of AWG wire sizes.

Size (AWG)	Area in Circular Mils	Area in mm^2
18	1620	0.821
16	2580	1.307
14	4110	2.083
12	6530	3.309
10	10380	5.260
8	16510	8.366
6	26240	13.296
4	41740	21.150
3	52620	26.663
2	66360	33.625
1	83690	42.406
1/0	105600	53.508
2/0	133100	67.443
3/0	167800	85.026
4/0	211600	107.219

Conductors larger than 4/0 are measured and sized directly in mils. A mil is a unit of wire measurement equal to one one-thousandth of an inch (0.001 inch). A circular mil is an area measurement equal to the area of a circle with a 0.001 inch (1 mil) diameter. A square mil is the area of a square that is 1 mil on a side. Figure 2–2 shows these two measurements.

Practical values for wire sizes fall into the thousands of circular mils category; for example, 250,000 circular mils or 250 kcmil. (Note: The preferred term for one thousand circular mils is kcmil. The older MCM is no longer used.)

Current Carrying Capacity (Ampacity)

Ampacity refers to the amount of current a conductor can carry without exceeding an allowable operating temperature. Excessive temperature rise can result in several negative consequences, including:

- Conductor melting (fusing)
- Insulation degradation and failure
- Fire

Ampacity is influenced by many factors, such as conductor material and size, ambient temperature, installation method, number of conductors in close proximity, and actual current levels.

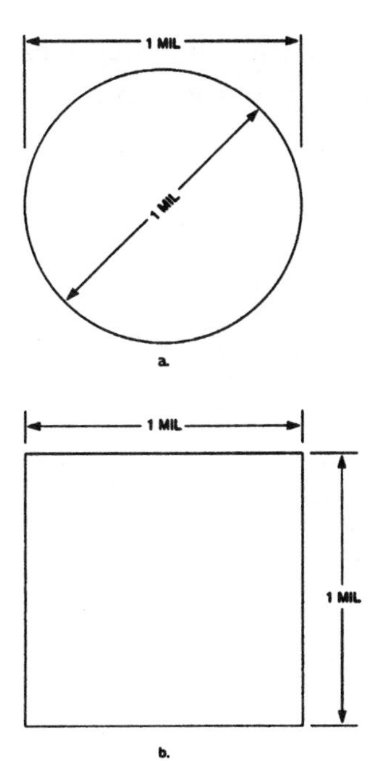

Figure 2–2 Conductor sizing: a. circular mil (CM) and b. square mil (SM).

Ampacity tables are available to determine what size cable may be used for a given application. Refer to the current edition of the *NEC*® for cable ampacity tables.

ELECTRICAL INSULATION

Classifications

Insulation can be classified by its application in various environments. For example:

- Chemical-resistant insulation is classified to corrosive and acid environments.
- Heat-resistant insulation is classified in terms of use in extreme or abnormal ambient temperatures.
- Moisture-resistant insulation is classified for use in areas where moisture is constantly present.
- Oil-resistant cable insulation is classified for use where oil is present in unusual or damaging conditions.

Table 310–13 in the 1990 *NEC*® lists all the types of insulation and identifies the trade names, maximum operating temperature, application provisions, sizes of conductors, thickness of insulation in units, and outer covering, if any.

Rubber

Rubber is one of the most common types of insulation. To prevent chemical reaction, it is usually separated from the conductor by various coatings and materials. Rubber is resistant to moisture and is not well suited for use with high temperatures (above 205° Celsius) or high voltages.

Various *NEC*® grades, such as Types R, RH, and RHW, are applied to rubber insulating materials to withstand high moisture or high temperature conditions and other abnormal application conditions. Rubber is commonly used for appliance cords, portable cords, conductor leads, and portable cables.

Elastomers

Elastomers are materials that can be compressed, stretched, or deformed and still snap back to their original shapes. In this respect they are like rubber. Four elastomers in fairly common use in modern insulation include Neoprene, Hypalon, Ethylene Propylene Rubber (EPR), and Styrene Butadiene Rubber (SBR).

Neoprene. Neoprene is one of the most important elastomeric materials in use. It has been used successfully as a cable jacket for about fifty years. Where service conditions are unusually abrasive, neoprene is highly recommended. In extreme environments, the neoprene construction is often reinforced with additional layers of cotton or other reinforcing members midway through the jacket wall. Not inherently flame retardant, neoprene jacketed cables must be compounded with the necessary flame-retarding chemicals. Like the other members of the elastomeric family, neoprene jackets are usually more expensive both to produce and purchase.

Hypalon (Chlorosulfonated Polyethylene). Hypalon possesses the majority of the properties of neoprene, plus it exhibits better thermal stability and resistance to both ozone and oxidation. Like neoprene, it is a fairly expensive material and requires continuous vulcanization (CV) for its fabrication.

Ethylene Propylene Rubber (EPR). Ethylene propylene rubber has been used in power cables for quite some time and has just recently been employed in fabrication of telecommunication and other types of wire and cable. When used as a jacketing material, EPR exhibits excellent weathering properties and is quite resistant to ozone. Unlike neoprene, EPR can be compounded to exhibit excellent electrical characteristics. Although EPR possesses good chemical and mechanical properties, it is not inherently flame retardant. Therefore, as in the case of some other jacketing materials, it must be compounded with flame-retarding additives.

Styrene Butadiene Rubber (SBR). This is a general purpose synthetic rubber characterized by relatively low tensile strength when stretched. In order for this material to be suitable for use as insulation, the base elastomeric material must be judiciously compounded with other materials, such as carbon black. The SBR materials are known for their excellent abrasion resistance and reasonably good electrical characteristics. However, when exposed to ozone, oil, or weather, they readily decompose. These compounds have a useful operating temperature range of about –55° Celsius to +90° Celsius. The material is highly flammable unless steps are taken to compound the base materials with adequate flame retardants.

Thermoplastics

Thermoplastic is another commonly used type of insulation that is much stronger than rubber. It is often used for conductor insulation and protective covering in cables rated below 2,000 volts. Thermoplastic materials can be softened by heat and flow under mechanical pressure, and they maintain their altered shape when cooled. Thermoplastic materials are more rigid than elastomers, have better electrical properties, and are lower in cost. They are also lightweight. The types of thermoplastic used primarily in insulation are described in the following paragraphs.

Polyvinyl Chloride (PVC). PVC is the most popular of the thermoplastics. PVC is a strong, water-resistant substance that may be compounded with many materials. Some compounds improve its mechanical attributes, while others accentuate its electrical characteristics. PVC possesses a high resistance to most acids, alkalis, and oils and can be processed with flame inhibitors to pass most flame tests.

PVC materials have been compounded to perform well at temperatures as low as –55° Celsius and as high as +105° Celsius. Formulations are available that allow PVC materials to yield good service life when used outdoors.

Polyethylene (PE). Polyethylene is used whenever low line loss is required. It has excellent electrical properties and is highly resistant to water, ozone, and oxidation. It has, however, low resistance to flames, gasoline, and solvents.

Thermosetting

Thermosetting insulation is similar in appearance to thermoplastic, but is somewhat thicker. This insulation is cross-linked polyethylene (XLP). It exhibits good resistance to heat, chemicals, moisture, and ozone. It does not soften in normal operating conditions, but will distort at temperatures above 90° Celsius. This insulation is used in high-voltage applications, such as motor leads, where temperatures do not exceed 90° Celsius, and in high-voltage cables.

Nylon

Nylon is a member of the polyamide class of resins and was originally developed for use as filament strands and synthetic fibers.

While nylon is not generally employed as a primary insulation, it is used as a secondary insulation and jacketing material where a high degree of abrasion resistance and/or heat and chemical resistance is desirable. Typical thermal resistance of an insulated conductor sheathed with nylon is 105° Celsius.

Nylon is not used as much as other insulation types because it has a strong affinity for moisture and has relatively poor electrical characteristics.

Teflon

Teflon®[1] is used as an insulation where high temperatures, varied chemicals, and moisture conditions exist. It also resists the destructive effects of oils. Teflon is used in temperature applications of up to about 200° Celsius. It is used with wiring on appliances, boiler controls, fire alarms, transformer leads, and motor leads.

Mineral Insulation

Mineral insulated cable is made of highly compressed magnesium oxide and covered with an outer metallic sheath, usually copper or alloy steel. Mineral insulation is designed to be as noncombustible as possible. It is used in feeders, services, and branch circuits in wet or dry locations.

Paper

Paper by itself is a mediocre insulation. However, when impregnated or immersed in other materials, paper insulation provides the best electrical characteristics of all currently used insulations.

1 Teflon® is a trademark of Dupont Corporation.

Solid. Solid paper insulation consists of layers of paper tape impregnated with mineral insulating oil. A tightly fitting lead sheath covers the insulation system and provides the best mechanical and chemical protection. Lead-covered solid paper insulation is being phased out in favor of newer, less expensive, and more easily used insulation systems.

Gas-Filled. This cable is constructed in a manner similar to solid paper insulation. Before the cable is leaded, the oil is drained into a nitrogen atmosphere. Gas feed channels are built into the cable to give the nitrogen free access to the conductors.

Oil-Filled. This cable is similar to solid insulation, except that the oil used is a relatively thin liquid, which is fluid at all operating temperatures.

Varnished Cambric

Varnished cambric is a smooth, yellowish-brown insulation that is being replaced by more modern types. It is composed of cotton cloth coated with insulating varnish, which is fabricated into tape. An oil compound is applied for waterproofing and to prevent friction between the insulation layers when the cable is bent.

Varnished cambric insulation has high dielectric strength, but breaks down when the temperature is above 85° Celsius. It was used in high-voltage, high-temperature applications such as generator leads, transformer leads, and substations. It is not frequently used in newer installations.

Polypropylene

Polypropylene is similar in many respects to polyethylene. It is used primarily as an insulating material and is typically harder than polyethylene. It is not as commonly used as polyethylene and finds application primarily as a thin wall type of insulation.

Silicon

Silicon is a very soft insulation with a relatively high temperature range. It has relatively good weather resistance, low moisture absorption, and is ozone and radiation resistant. It has low mechanical strength and scuffs easily and is used primarily for insulation as opposed to jacket applications.

Tefzel

Tefzel®[2] has excellent electrical properties; good chemical, heat, radiation, and flame resistance; and is quite tough. Tefzel is a fluorocopolymer thermoplastic material.

Halar

Halar®[3] is a thermoplastic fluoropolymer material. It has excellent electrical and chemical properties. It also exhibits good physical toughness to heat, radiation, and flame.

JACKETS AND SHIELDS

Many cables are covered with protective jackets for additional thermal, chemical, mechanical, and environmental protection. If the cable has a metal shield (see construction details later in this chapter), the jacket is usually applied directly over the shield. Otherwise, it is placed directly on the insulation.

Most of the insulation materials already discussed may be used as cable jackets. The most popular jacketing materials employed include PVC, PE, neoprene, hypalon, and EPR. Polyurethane and thermoplastic elastomer (TPE) jackets are also gaining popularity.

Polyvinyl Chloride (PVC)

PVC is the most popular jacketing material. PVC resins may be compounded in many ways to afford normal mechanical protection in cables suited for indoor applications. Other vinyl compounds find widespread application in the area of process control cables. Process control cables require extremely rugged construction to withstand harsh operating conditions. PVC jackets are compounded with other chemicals for indoor or outdoor use in a wide range of temperatures: from –60° Celsius to +105° Celsius. The rugged PVC jackets used in instrumentation cables are also compounded with flame-retardant chemicals so that they have an extra measure of flame retardancy. Because of the additional compounding requirement, this type of PVC jacket is generally more expensive than those used in the more general-purpose PVC jacketed wires and cables.

2 Tefzel® is a trademark of the Dupont Corporation.
3 Halar® is a trademark of the Ausimont Corporation.

PVC jackets exhibit relatively good resistance to abrasion and the absorption of water. However, in places where constant standing water is expected, or where there are regulatory agency requirements with which the cable must comply, PVC jacketing is not preferred.

Polyethylene (PE)

PE cable jackets are preferred over PVC jackets where constant water conditions are expected, because PE is much more impervious. PE materials are not naturally flame retardant, but they can be made so by compounding with additives. Even when compounded, they are not as resistant to fire as PVC. Typical service life for polyethylene jacketed cable is over twenty years.

Polyurethane

Polyurethanes possess excellent abrasion resistance and excellent low temperature flexibility. This family of compounds may be extruded by conventional means without vulcanizing. Polyurethanes are hygroscopic and flammable.

Thermoplastic Elastomers (TPE)

The TPE family of materials has found increasing application as both a primary insulating material and a jacketing material. Thermoplastic elastomer materials possess excellent environmental and chemical characteristics, which make them a natural for jacketing applications.

Flamarrest

Flammarrest®[4] is a plenum-grade, chloride-based jacketing material. It has low smoke- and flame-spread properties. Flammarrest jacketed cables meet UL Standard 910, Plenum Cable Flame Test.

Metal-Clad Cable

Metal-clad cable consists of one or more separately insulated conductors twisted together with cords of jute that run the full length of the cable. The conductor assembly is then wrapped with a binding tape or enclosed in a thermoplastic jacket. Next, the entire assembly is enclosed in a metallic sheath. The sheath may be

4 Flamarrest® is a registered trademark of Dupont Corporation.

smooth and seamless, welded and corrugated, or interlocking. The interlocking sheath is in the form of interlocking metal tape armor similar to Type AC armored cable.

There are three types of cable in this category: MC, ALS, and CS. These types differ in their construction and in the material used for the metal jacket that encloses the cable.

Metal-clad cables are designed for service voltages of up to 15,000 volts. Cables designed for 600 volt service have one or more stranded grounding conductors running their full length.

Type MC. This type is enclosed in a corrugated metallic sheath or in interlocking metal tape made of galvanized steel, aluminum, or bronze. The metallic sheath may have a plastic covering to protect it from corrosive conditions.

Type ALS. Type ALS metal-clad cable is enclosed in a smooth, seamless aluminum sheath.

Type CS. This jacket is composed of a smooth, seamless copper or bronze sheath.

Lead Sheath

Lead sheath is used where cable will be buried directly in the ground. Lead is also used where cables will be subjected to water, corrosion, or abrasion. It is sometimes used with an armored jacket.

ELECTRICAL POWER CABLE CONSTRUCTION

Low Voltage (Less Than 2,000 Volts)

Figure 2–3 shows the general construction features of low-voltage cable. The center conductor is surrounded by an insulation layer, which provides the electrical insulation. The insulation may be covered by a protective jacket. Note that while the protective jacket may have some electrical insulating ability, the actual insulation characteristics of the cable are vested completely in the insulation layer.

The materials used in this system were described previously in this chapter.

Medium Voltage (2,001 Volts–35,000 Volts)

Because of the greater electrical stress, medium-voltage cable is more complex in construction than low-voltage cable. Medium-voltage cable is composed of five

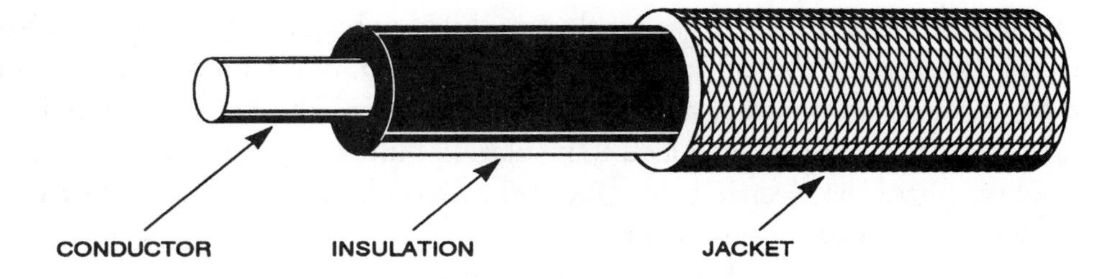

Figure 2–3 Low-voltage cable construction.
(Courtesy of Cadick Corporation)

or more individual layers, each of which has a specific purpose. The following explains the six most commonly used layers. Each of the layers is diagrammed in Figure 2–4.

Conductor: Made of copper, aluminum, or other conducting material, this layer is responsible for the actual conduction of the electrical current.

Semiconductor: This layer is also called the "strand shield" or "strand screen." Air voids between the conductor and the insulation allow the buildup of very high-voltage gradients. The strand screen "short circuits" these voltage gradients and prevents the formation of corona.

Insulation: This is the electrical insulation layer. As with low-voltage cable, 100 percent of the electrical insulating capability of the cable is vested in the insulation layer.

Semiconductor: This layer is similar to the strand shield and serves a similar purpose. This outer semiconducting shield also helps to enclose completely and equalize the electric field within the cable. This will be explained in greater detail shortly.

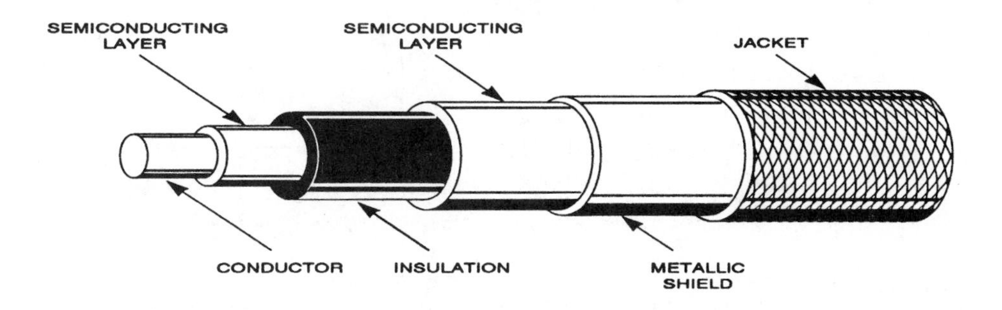

Figure 2–4 Medium-voltage cable construction.
(Courtesy of Cadick Corporation)

Metallic Shield: This layer serves as an electric shield, which contains and equalizes the electric field inside the cable. Consider Figure 2–5. With no shield, the field would be unevenly distributed, as shown in Figure 2–5a. This causes unequal and excessive electrical stress and can lead to insulation failure. Figure 2–5b shows how the shield equalizes the electric field. Under these conditions, the cable life will be greatly extended. Note that this is not usually a problem at low voltages. Therefore, cable for low voltages is not typically designed in this manner.

Jacket: The jacket provides environmental protection for the cable. Chemical, ultraviolet, and moisture damage is kept from the insulation layer by the jacket. The jacket also protects against mechanical damage.

ELECTRICAL COMMUNICATIONS CABLE CONSTRUCTION

Coaxial Cable

Construction. Coaxial cable is a type of communications cable developed during World War II. It was originally designed to transmit radio frequency signals from transmitters to antennas and/or from antennas to transmitters. It replaced older types of cables, such as parallel twin-conductor, and was generally superior to the older methods in terms of ease of installation, ruggedness, and general utility. Typical layers are shown in Figure 2–6. Coaxial cable is the lowest-loss copper conductor cable in use for communications systems. Only fiber optic cable offers superior attenuation characteristics.

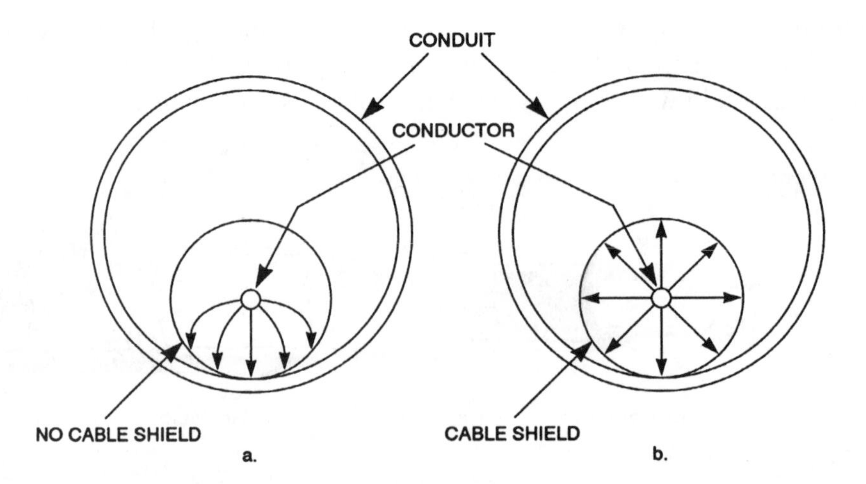

Figure 2–5 Electric fields inside cables: a. without shield and b. with shield.
(Courtesy of Cadick Corporation)

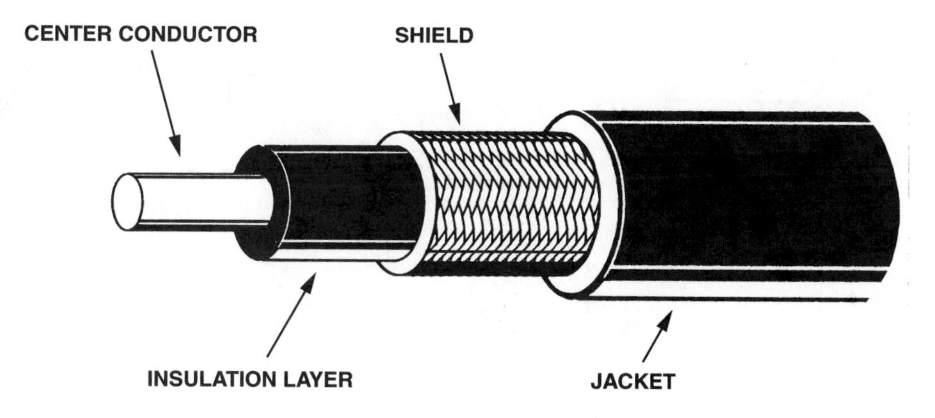

Figure 2–6 Coaxial cable construction.

Center Conductor: Usually made of copper or aluminum, the center conductor carries the actual signal.

Insulation Layer: The insulation (also called the dielectric) may be one of the many types of synthetic material, such as polyethylene solid or foam. The insulation layer is surprisingly thick on coaxial cables used for higher power work because of the relatively high voltages that can build up during transmission. Coaxial cable intended for very high-frequency work may use air for the insulator, and the center conductor may be held in place by doughnut-shaped insulating wafers.

Shield: The shield, or braid as it is sometimes called, is used for two important functions. First, it serves as the return path for the signal current. Perhaps even more important, however, is its second use, which is to "shield" the signal conductor from the influence of external electromagnetic fields. It also helps to contain the electromagnetic field set up by the current flow in the center conductor. This action is especially important at the high frequencies used by the cable television industry. Coaxial cables may be double shielded if they are intended for high security or high reliability applications.

Jacket: The outer jacket protects the cable from mechanical or environmental damage. It may be made of a thermoplastic material such as PVC or, in high-exposure environments (such as cable TV installations), it may be made of an aluminum metal sheath. Some coaxial cables use a metal jacket at the shield.

Other constructions of coaxial cable are also available for different types of applications. Two of these types, shown in Figure 2–7, are the twinaxial and triaxial cable. Twinaxial cable is used in certain types of computer networks. Triaxial cable was among the first types of coaxial cable to be used in local area networks (LANs).

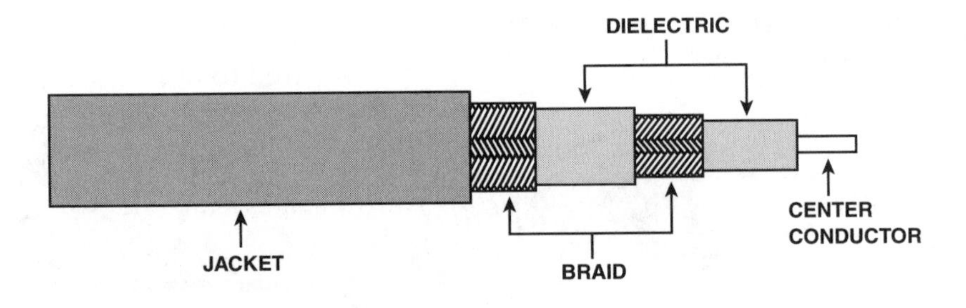

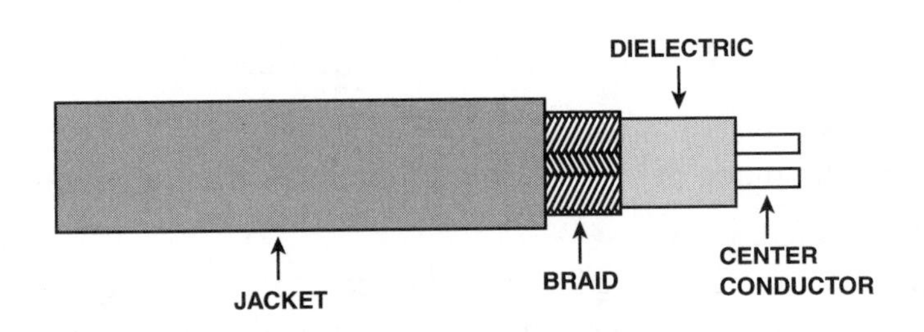

Figure 2–7 Twinax and triax cable.

Characteristic Impedance. Conductors behave differently at very high frequencies. Consider, for example, a piece of wire that is 45 feet long. At 60 Hz power system frequencies, the wire will pass all of the energy from one end to the other end—the generator to the load. At 10 MHz (10,000,000 Hz), however, the piece of wire looks like an antenna. As the 10 MHz signal is transmitted down the wire, a significant portion can radiate off into space, thereby losing a great deal of signal strength and information. Note that 10 MHz is the approximate operating frequency of a 10BaseT computer network.

A very important electrical communications cable characteristic is the characteristic impedance. The characteristic impedance of the cable must match the impedance of all other components in the system. Transmitters, couplers, splices, connectors, and receivers must all be of the same characteristic impedance or system losses will increase greatly.

Most computer networks use cables with 50 ohm characteristic impedance. Cables of this type are the RG-8/U and RG-58/U family. Cable TV systems use primarily 75 ohm cables. The RG-6/U, RG-59/U, and RG-11/U type coaxial cables have 75 ohm characteristic impedances.

Attenuation. Attenuation is the loss of power that a signal experiences as it moves down a cable. Coaxial cable has a very low loss compared to other types, as shown in Table 2–3.

Table 2–3 Loss comparison for various types of communications cables.

Cable Type	dB Loss per 100 Meters (10 MHz)	db Loss per 100 Meters (100 MHz)
RG-59/U	3.6	11.2
RG-6/U	2.6	8.9
RG-11/U	2.2	6.6
RG-58/U	3.6	12.5
RG-8/U	1.8	6.2
Cat 3 UTP	9.7	xxx
Cat 4 UTP	6.9	xxx
Cat 5 UTP	6.5	22

Note that 3 dB equals a ½ power loss. Also note that the figures given are for general comparison and may vary by manufacturer.

Twisted-Pair Cables Basics

The simplest form of a twisted-pair cable is shown in Figure 2–8. This type of cable is used almost universally for communications circuits, from simple telephone circuits to high-speed computer networks.

Twisted-pair cables minimize noise and cross-talk in two basic ways. First, the twists keep the two wires very close together and thereby almost eliminate the net electromagnetic field created by the pair. This greatly reduces the cable's tendency to induce signals in adjacent conductors.

Second, the signal applied to such a cable is balanced. That is, the signal on each cable is equal in magnitude and opposite in sign to that on the other cable. The receiver at the end of the cable uses the difference between the two signals. Most noise that is coupled into the cable tends to be "common mode" noise. That is, it affects each conductor equally. Therefore, the noise is canceled at the receiver.

Figure 2–8 Twisted-pair cable.

Twisted-pair cables are manufactured in three common formats: unshielded (UTP), screened (sUTP, scTP, or FTP), and shielded (STP). While each type offers its own special characteristics, UTP remains the most popular and common type.

Unshielded Twisted-Pair Cable (UTP)

Starting originally as a single twisted pair of wires in an overall insulated covering, the most common form of UTP is now four individually jacketed pairs held together within a single outer jacket (see Figure 2–9). The wire size is usually #24 AWG, but #22 is sometimes used.

Note that 25-pair UTP is also commonly used for vertical backbone wiring.

UTP is rated by frequency in five basic categories, referred to as Cat 1 through Cat 5:

Category 1: Not frequency rated, although it is still used in some analog telephone applications
Category 2: 64 kHz
Category 3: 16 MHz
Category 4: 20 MHz
Category 5: 100 MHz (note from Table 2–3 that the loss at this frequency is a very large 22 dB)

International standards organizations do not recommend Categories 1 and 2 for use. The lowest recommended category is Cat 3.

While Category 5 has the best overall frequency response and signal attenuation, it is also the most critical from the standpoint of installation. For example, the twists must continue to within at least ½ inch of a termination. Otherwise, the frequency response will be severely damaged.

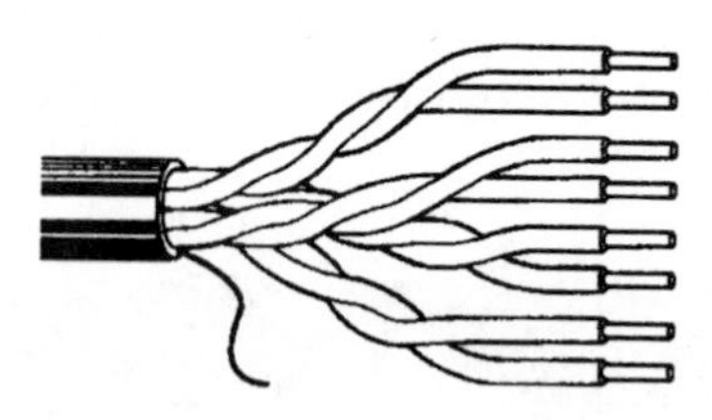

Figure 2–9 Four-pair unshielded twisted-pair cable.
(Courtesy of Belden Wire and Cable Company)

Screened Twisted-Pair Cable (sUTP, scTP, or FTP)

Screened twisted-pair cable is shielded UTP cable. It is very similar to UTP except that it has a foil shield that completely covers the four pairs inside. It differs from STP in that each pair is not individually shielded. scTP has the same basic electrical characteristics as UTP, plus an increased immunity to outside noise. Figure 2–10 shows typical scTP construction.

Shielded Twisted-Pair Cable (STP)

Shielded twisted-pair cable, Figure 2–11, was first made popular by IBM for its Token Ring network system. Although STP offers higher useful frequency coverage, UTP remains more popular because it is much easier to work with.

Length and Thermal Performance Characteristics

General performance characteristics and maximum lengths for twisted-pair cables and fiber cables are shown in Table 2–4. Table 2–5 lists the thermal characteristics.

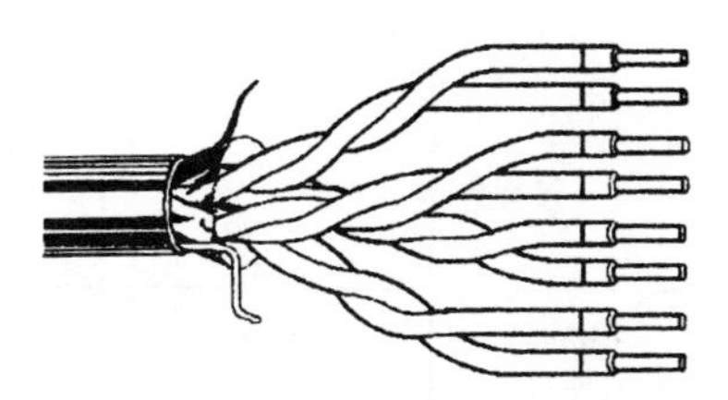

Figure 2–10 Screened UTP.
(Courtesy of Belden Wire and Cable Company)

Table 2–4 Backbone cable distances.

	Backbone Cables: Main Cross-Connect to Telecommunications Room	
Cable	**Frequency**	**Distance (m)**
UTP	Voice grade	800
Cat 3 UTP	5–16 MHz	90
Cat 4 UTP	10–20 MHz	90
Cat 5 UTP	20–100 MHz	90
STP	20–300 MHz	90
62.5/125 Fiber	—	2,000
Single-Mode Fiber	—	3,000

Table 2–5 Cable classifications and flammability ratings.

Cable Type	Classification		
	Plenum	**Riser**	**Commercial**
Communication	CMP	CMR	CM
Fiber with no metallic conductors*	OFNP	OFNR	OFN
Fiber with metallic conductors*	OFCP	OFCR	OFC

*Metallic conductors refer to any part of the cable that can conduct electricity, including strength members. The metallic conductor does not have to be specifically intended to carry electricity or electrical signals. See Chapter 3 for more information about fiber optic cables.

Name	Type	Impedance (ohms)	Description
Type 1/1A 	STP	150	Two individually shielded, solid-conductor, 22 AWG twisted pairs surrounded by an outer braid shield. Both pairs are suited for data transmission.
Type 2/2A 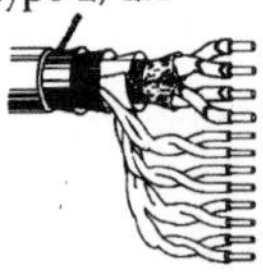	STP/UTP	150 (STP)	Two solid-conductor, 22 AWG twisted pairs surrounded by an outer braid shield. Four solid-conductor, 22 AWG twisted pairs outside the braid for telephone use.
Type 3 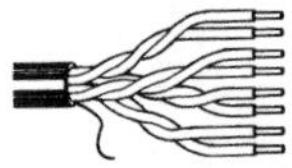	UTP	100	Four solid-conductor 24 AWG twisted pairs, unshielded. Similar to Category 2; not recommended for network applications.
Type 6/6A 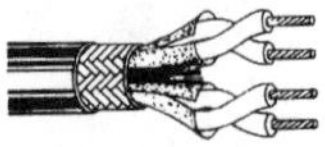	STP	150	Two stranded-conductor, 26 AWG twisted pairs surrounded by an outer braid. Similar to Type 1 except for wire gauge. Used for short patch cords.
Type 9/9A 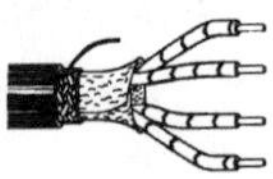	STP	150	A plenum-rated cable with two 26 AWG twisted pairs surrounded by an outer shield.

Figure 2–11 Shielded twisted-pair cabling system.
(Courtesy of Belden Wire and Cable Company)

FIBER OPTIC CABLE CONSTRUCTION

Construction of Fiber Optics

As described earlier in this chapter, optical fibers are made of glass or plastic with at least two concentric layers. The innermost layer, called the core, serves as the main conduit for the light. While the core can be, and sometimes is, made of plastic, cables suitable for network applications are almost universally made of glass.

The next layer, the cladding, reflects the light waves that try to escape from the core. The outside layer, called the buffer, is used primarily for mechanical protection.

Even with a buffer, the single fiber is not strong enough to be used without additional mechanical protection. This is especially important during installation. Figure 2–12 shows the different ways in which fibers are constructed in cables for network service.

Characteristics of Fiber Optics

Light travels down a fiber by reflecting from the interface between the core and the cladding (see Figure 1–4). A light beam that impacts the interface at an

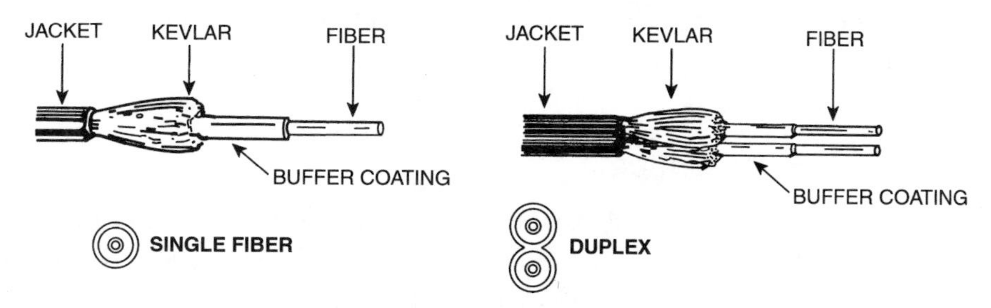

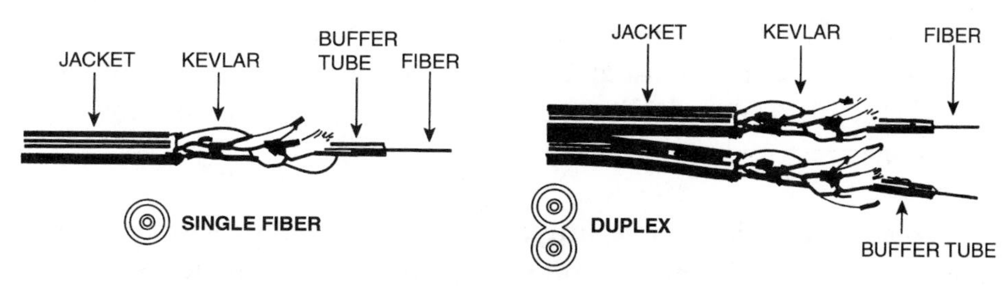

Figure 2–12 Common cable constructions.
(Courtesy of Belden Wire and Cable Company)

angle less than the critical angle will be totally reflected back into the core. Therefore, the light beam will be continually reflected from one side of the core to the other, down the length of the fiber. The only loss that occurs is the normal attenuation of the light moving through the glass core.

Depending on how the fiber is constructed and on the frequency of the light wave, there may be only one or many tens of thousands of paths for the light beams as they move. Each path is called a mode. If one frequency of light with no information bits is transmitted down the cable, there will be virtually no bandwidth to the signal. As information is added to the light wave, more bits per second will move down the fiber. This higher data rate will result in a wider spread of frequencies.

In fiber, the bandwidth is limited by an action called dispersion. Dispersion may be created by material losses or by multiple modes arriving at the same time and interfering with each other. Ultimately, then, a single mode fiber has much greater bandwidth than a multimode fiber. This means that the single mode fiber can transmit substantially more information than a multimode.

Note that the bandwidth of fiber is also substantially greater than for copper communications cable. First, the operating frequency of fiber is much, much greater than for copper. This means that the deviation from the center frequency is more limited for copper, simply because it does not have as far to move. Second, the change in attenuation versus frequency is much greater for copper than for optical fiber (see Table 2–3).

Advantages of Fiber Optics

Information Capacity. Fiber can carry substantially more information than any other type of cable. For example, the bandwidth of the 62.5/125-μm fiber[5] is between 1.5 Gbps and 5 Gbps at 100 meters.[6] The bandwidth of a Category 5 cable is only 100 MHz over the same 100 meters.

Low Loss. As evidenced by Table 2–3, optical fiber has significantly lower loss than does copper. The useful distance of copper is meters or, at most, hundreds of meters in a network. The useful distance of fiber, because of its much lower attenuation, is kilometers.

Electromagnetic Immunity and Security. Despite the shields, twists, and magnetic field cancellations, electric cable is still subject to electromagnetic disturbances. In fact, a very high percentage of network outages are caused by electromagnetic

5 The core diameter is 62.5 μm and the cladding diameter is 125 μm.
6 1 Gbps is equal to 1 billion bps.

interference and noise. Optical fiber does not have this problem because it operates at frequencies well above normal electromagnetic values.

Also, fiber does not emit an electromagnetic field and cannot be eavesdropped by remote receivers. Also, tapping a fiber optic cable is quite difficult. These two characteristics taken together make fiber optic cable among the most secure of all communications channels.

Weight and Size. Fiber cable, even multiple fiber cable (see Figure 2–13), is substantially lighter than electrical cable. When this is calculated based on bandwidth per pound, fiber is even better. Fiber is also much smaller and therefore easier to pull into small areas.

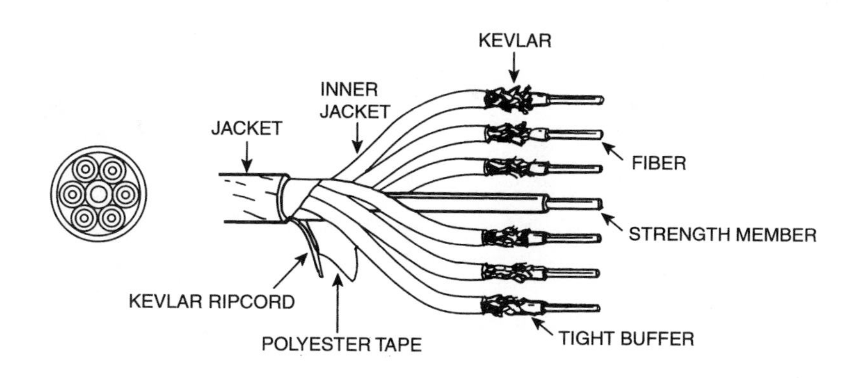

Figure 2–13 Multifiber breakout cables.
(Courtesy of Belden Wire and Cable Company)

Safety. Even though signal voltages are usually small, a chance of shock does exist. This is especially true if the cable contacts electrical power lines. Fiber optic cable, in contrast, does not conduct electricity and represents no electrical shock hazard.

Chapter 3

Cable Installation, Splicing, and Termination

KEY POINTS

- How can wire and cable be installed?
- What are the advantages and disadvantages of the various installation methods?
- How are wire and cable spliced and terminated?
- What are the differences between medium-voltage and low-voltage splicing and termination?

INTRODUCTION

Wire and cable are subject to constant attack from electrical, mechanical, chemical, and environmental elements. While the use of proper materials is the key starting point to the long-term survival of wire and cable, proper installation is also a major factor. The best equipment will not survive if it is not installed, spliced, and/or terminated properly. The type of wiring installation depends upon the environment, whether it is to be aboveground or underground, the temperature, and the conductor's size and type. Figure 3–1 shows the various types of installations and their applicabilities to various situations.

The *NEC®* and other standards have many regulations governing acceptable installation for wire and cable. Considerations such as moisture, temperature, and mechanical stress are considered. This chapter will review many of these items, as well as cover some workmanship concerns.

NEC® ARTICLE	PERMITTED LOCATIONS: INSIDE BUILDINGS	OUTSIDE BUILDINGS	UNDERGROUND	CINDER FILL	EMBEDDED IN CONCRETE	WET LOCATIONS	DRY LOCATIONS	CORROSIVE LOCATIONS	SEVERE CORROSIVE LOCATIONS	HAZARDOUS LOCATIONS	MECHANICAL INJURY	SEVERE MECHANICAL INJURY	EXPOSED WORK	CONCEALED WORK	SERVICES OUTSIDE
346 RIGID STEEL CONDUIT, ZINC-COATED	P	P	P	E	P	P	P	P	P	P	P	P	P	P	P
346 RIGID STEEL CONDUIT, ENAMELED	P	X	X	X	E	X	P	P	X	P	P	P	P	P	X
348 ELECTRICAL METALLIC TUBING	P	P	P	E	P	P	P	E	P	E	P	X	P	P	P
350 FLEXIBLE STEEL CONDUIT	P	E	E	E	E	E	P	E	X	E	P	X	P	P	E
354 UNDERFLOOR STEEL RACEWAYS	P	–	–	–	P	–	P	X	X	X	–	–	X	P	–
356 CELLULAR STEEL RACEWAYS	P	–	–	–	P	–	P	X	X	X	–	–	–	P	–
362 STEEL WIREWAYS	P	E	X	X	X	P	P	X	X	X	P	X	P	X	X
364 STEEL BUSWAYS	P	E	X	X	X	E	P	X	X	X	P	X	P	X	E
351 LIQUIDTIGHT FLEX, CONDUIT	LIMITED TO USE FOR CONNECTION OF PORTABLE EQUIPMENT OR FOR MOTORS IN WHICH FLEXIBILITY IS NEEDED.														
374 AUXILIARY STEEL GUTTERS	LIMITED TO USE AS SUPPLEMENTAL WIRING SPACE FOR SERVICES, ETC.														

P—PERMITTED E—EXCEPTION X—NOT PERMITTED – —CONDITIONS DO NOT APPLY

Figure 3–1 Types of installations.

TERMS

The following terms and phrases are used in this chapter.

Bending Radius: When a cable is routed around a corner, it follows a path that is a quarter circle. The radius of this quarter circle is called the bending radius. See Figure 3–21.

Conduit: A tube or trough for protecting electrical wires and cables. It may be a solid or flexible tube in which insulated electrical wires are run.

Fish Tape: A long, flexible tape used to thread and pull cable through a conduit.

Penciling: Tapering insulation into the shape of a pencil. This is normally done as part of a splicing or terminating procedure.

Splice: The physical connection of two or more conductors to provide electrical continuity.

Termination: The method used to prepare a cable or wire for connection to other equipment. A proper termination allows for uninhibited current flow and adequate insulation. A termination differs from a splice in that the other equipment to which the wire or cable connects is *not* a continuation of the wire or cable.

UNDERGROUND CONDUIT SYSTEMS

Underground conduit systems are also known as duct systems and consist of one or more conduits spaced closely together. These systems are often buried directly in the ground in concrete casing. Figure 3–2 shows the general layout for such a system.

The concrete casing is buried in a trench below the earth's surface. The depth of the conduit bank is determined by the voltage and type of service. Duct lines terminate in underground vaults called manholes. Conduits may be made of several materials, including fiber, polyvinyl chloride (PVC), vitrified tile, polyethylene, styrene, and monolithic concrete. The following paragraphs will discuss the two most commonly used materials, fiber and PVC.

Fiber Conduit

Fiber conduit is made of a wood pulp impregnated with a bituminous pitch. The different sections of the conduit are connected by self-aligning joints. Fiber

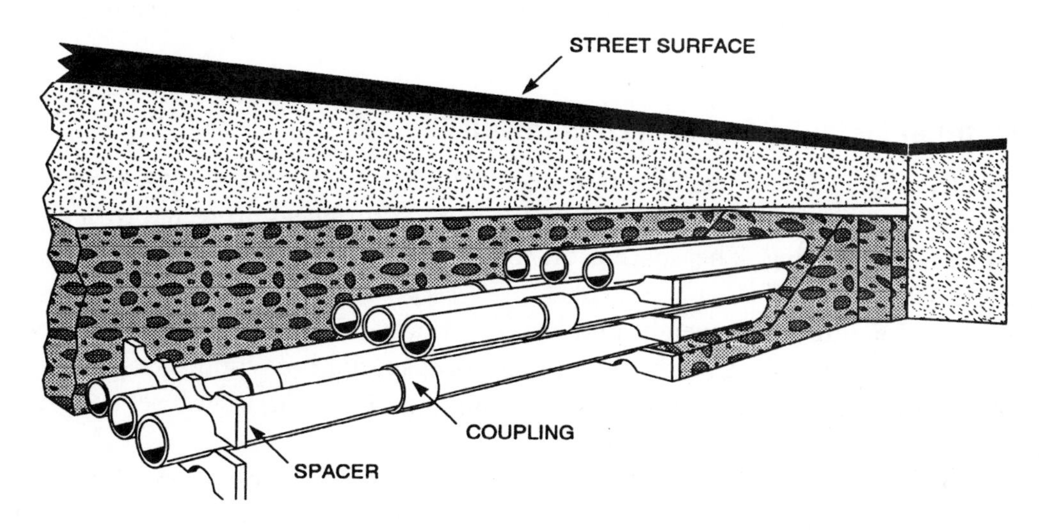

Figure 3–2 Underground conduit system.

conduit may be either Type 1 or Type 2. Type 1 must be encased in concrete while Type 2 may be buried directly.

Fiber conduit is resistant to corrosion and chemicals such as acids or alkalis. It has a high crushing and tensile strength and is impervious to moisture. Fiber conduit is not suitable for exposed installations.

Polyvinyl Chloride (PVC)

Several plastic materials are successfully used for conduit, including PVC, polyethylene, and styrene. Of these, PVC has proven to be the most popular. PVC is low in cost, easy to install, and comes in lengths of up to 30 feet.

PVC is first installed in banks and tiers with appropriate spacers, and concrete is then poured around it. PVC may be used both underground and aboveground, depending upon the requirements of the installation.

ABOVEGROUND CONDUIT SYSTEMS

Metal or nonmetal conduits may be used in exposed or concealed locations aboveground. Such conduits are made in a variety of materials and configurations.

Preassembled Systems

A relatively new introduction to underground distributions systems is the so-called preassembled system, which has cable already inserted in the conduit. These "cable-in-conduit" systems allow simultaneous installation of cable and conduit and prevent the installer from having to pull the cable in after the conduit or duct has been installed.

Electrical Metal Tubing (EMT)/Thinwall

Electrical metal tubing has a thin steel wall; therefore, it is called thinwall conduit. The outer surface is galvanized, while the inner surface is protected with a zinc or an enamel coating. The coating on the inside provides a smooth surface for pulling wire or cable.

This conduit is lightweight and ranges from ½ inch to 4 inches in diameter. Thinwall conduit cannot be threaded; therefore, special box connectors and couplings must be used. The connectors and couplings are secured with set screws or compression denting or by force fit.

Thinwall conduit is vulnerable to corrosion and rust; however, special types are available with greater resistance to these conditions. Figure 3–3 compares thinwall conduit with rigid conduit. Figure 3–4 illustrates a variety of fittings used with EMT.

Intermediate Metal Conduit (IMC)

Intermediate metal conduit falls between EMT and rigid conduit in terms of weight, strength, and cost. It is made of aluminum or steel and is similar to rigid conduit in many ways. IMC can be threaded and provides a more rugged installation than EMT; however, it has a thinner wall than rigid conduit and should not be used if the installation is subject to severe mechanical stress.

Rigid Conduit

Rigid conduit (Figure 3–3) is the most commonly used of the metal conduit classes. Rigid conduit is available in sizes from ½ inch to 6 inches in diameter and is made of aluminum or steel. Rigid conduit is usually threaded for installation.

Galvanized Steel. This type is suitable for indoor or outdoor use and can also be embedded in concrete. Galvanized steel is easily threaded, very strong, and has good corrosion resistance. It is also relatively heavy and costly and corrodes easily if the galvanized coating is penetrated. Because galvanized steel is magnetic, it is a cause of magnetic power loss and poor heat dissipation. Galvanized rigid conduit is also called "white" conduit.

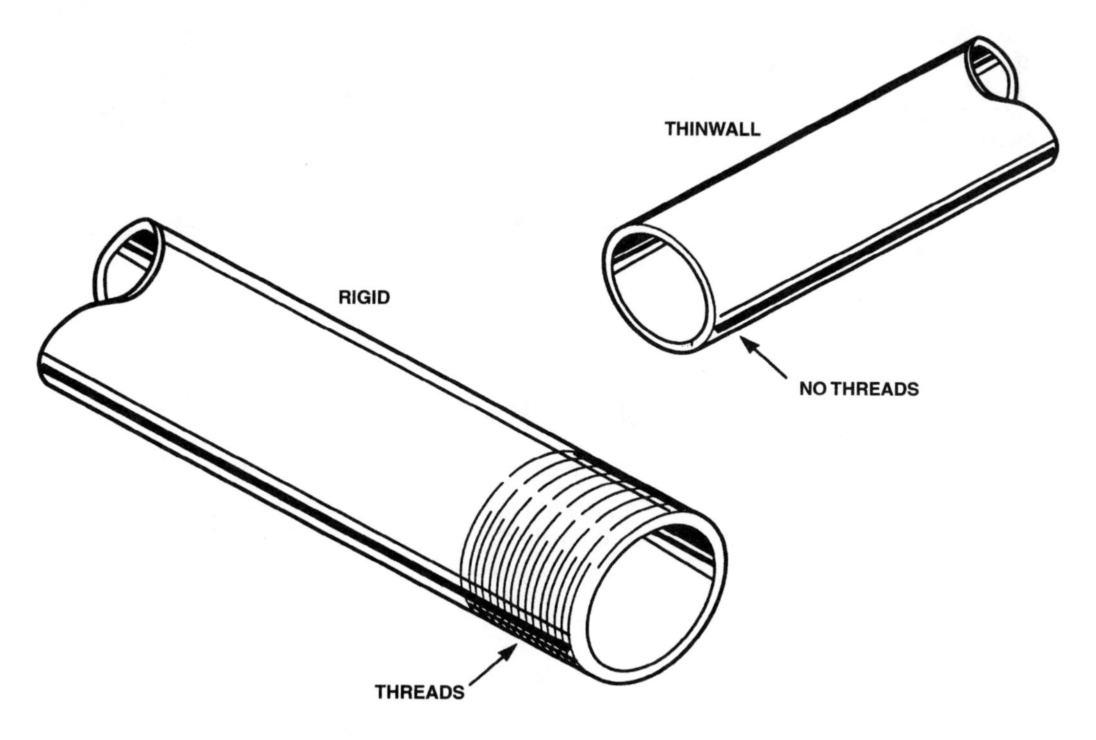

Figure 3–3 Conduit comparison.

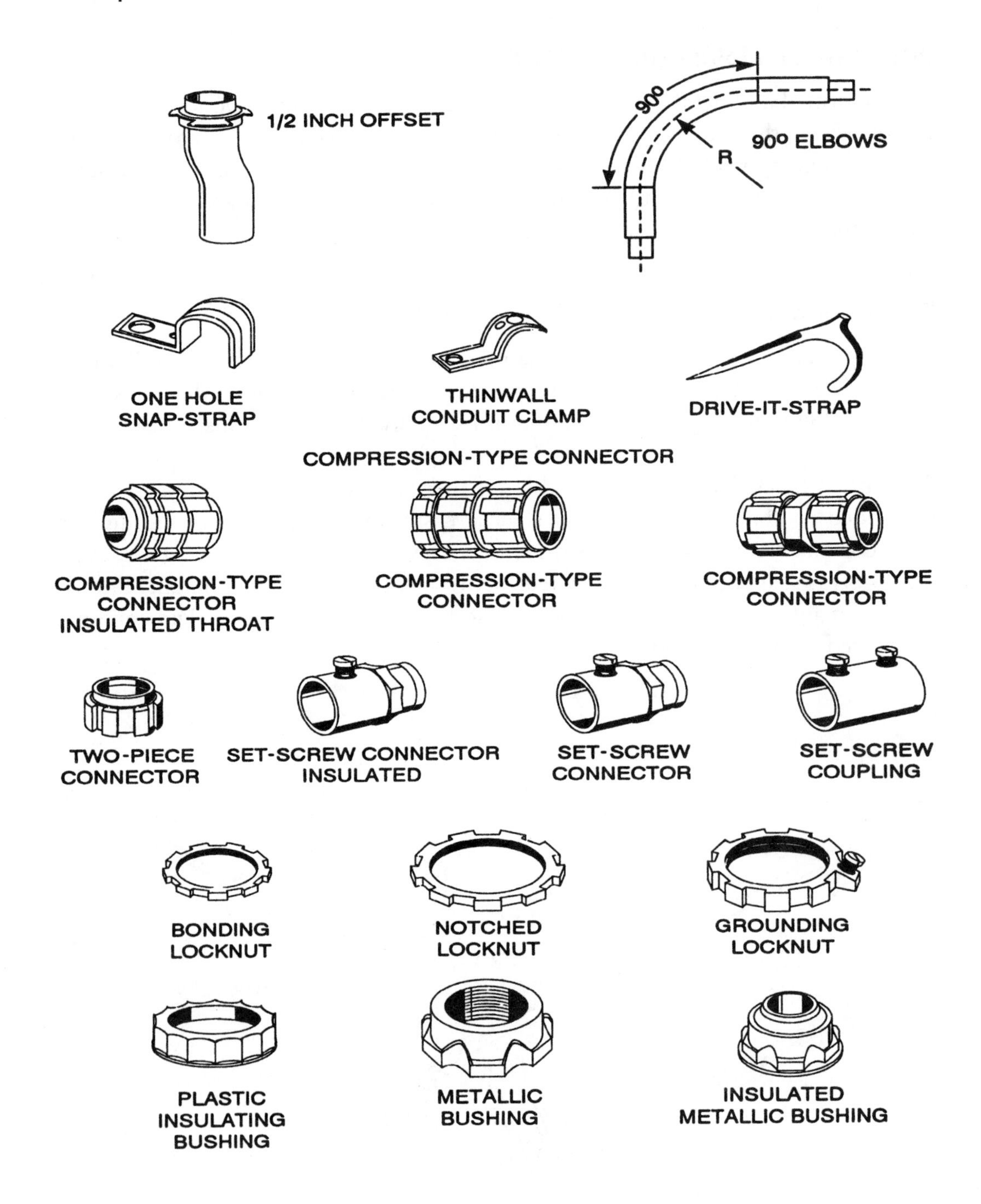

Figure 3–4 Thinwall conduit fittings.

Enameled Steel. Enameled steel (black) conduit is coated with a black enamel. This conduit is commonly used for general interior wiring.

Sheradized Steel. Sheradized steel (green) conduit has a special corrosion-resistant coating and is used in installations in which high levels of corrosion are expected.

Aluminum. Aluminum has been increasingly used for exposed indoor and outdoor locations. It is one-third the weight of galvanized steel and is noncorrosive in most atmospheres. Aluminum is not as strong as steel, corrodes in some soils, and reacts with concrete.

Polyvinyl Chloride (PVC)

PVC has a number of advantages and is used in both aboveground and underground systems. The inner walls are smooth and facilitate long cable pulls. The heavy wall type may be found in corrosive or chemical environments. PVC is nonporous and relatively impervious to moisture. PVC is not used in concealed spaces of combustible construction or to support fixtures or other equipment. This conduit is nonmagnetic and nonconducting.

PVC tends to be less expensive than other types of conduit and may be found in Type A thinwall, schedule 40 heavy wall, and schedule 80 and 120 extra heavy wall. Schedule 40 has an inside diameter similar to that of rigid metallic conduit. Sunlight-resistant types of PVC are also available.

PVC is coupled using bonding adhesives or glue. Sizes range from 1 inch to 6 inches in diameter. Ten-, twenty-, and thirty-foot lengths are available. PVC has relatively low mechanical strength and is susceptible to heat damage for temperatures above 120° Celsius.

Liquidtight Flexible Nonmetallic Conduit (LFNC)

LFNC is used where the extra flexibility and corrosion resistance of a nonmetallic conduit is required. LFNC is constructed in three different types:

- A smooth, seamless inner core covered with reinforcing layers bonded together.
- A one-piece assembly consisting of a smooth inner core with built-in reinforcement.
- A corrugated inner and outer surface with no reinforcement.

LFNC is made of PVC or some other similar waterproof, corrosion-resistant, insulating material.

Flexible Metal Conduit

Flexible metal conduit is made of a single strip of galvanized steel or aluminum wound into a longitudinal spiral and interlocked to form a continuous flexible tube. The resulting construction has many of the advantages of solid metal conduit, in addition to being lightweight and very easy to install. Lengths of up to 250 feet are available. Except for liquidtight conduit (discussed shortly), flexible metal conduit may not be used in wet environments unless the enclosed wire is rated for the location. Bending radii must be closely observed to prevent pinching the cable insulation.

When flexible metal conduit is enclosed in a plastic sheath, liquidtight flexible metal conduit is created. This type of conduit is especially useful for motor lead connections and other installations that require great flexibility and may be subject to water problems. Liquidtight conduit may not be used where physical damage is likely to occur. Figure 3–5 shows a cross section of liquidtight flexible metal conduit.

Wireways and Wire Ducts

Wireways are sheet-metal troughs with hinged or removable covers for housing and protecting wires and cables. These systems are often used when equipment and control devices are located close to each other.

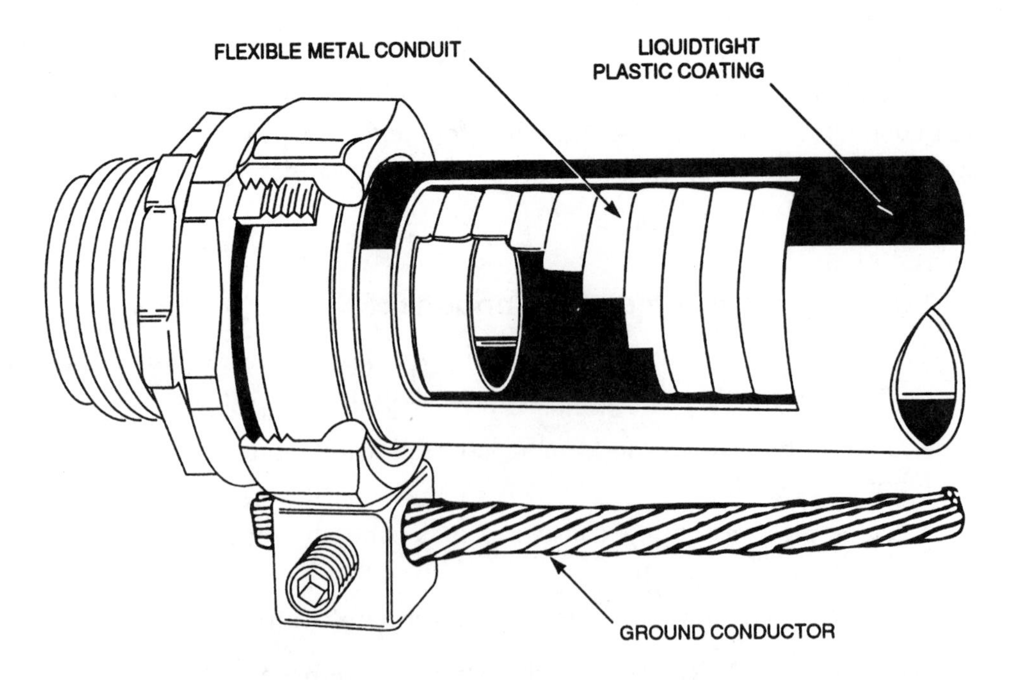

Figure 3–5 Liquidtight flexible conduit.

Cable Trays

A cable tray is an assembly of units made of metal or other noncombustible materials that form a continuous, rigid support for cables; see Figure 3–6. Cable trays are used throughout industry, and they greatly simplify the installation of wire and cable. There are four different types of cable trays.

Ladder Type. A ladder-type cable tray is a prefabricated metal structure consisting of two longitudinal side rails connected by individual transverse members, which provide the cable supporting means.

Trough Type. This is a prefabricated metal structure greater than 4 inches in width. It consists of a ventilated bottom and has closely spaced cable supports within integral or separate longitudinal side rails.

Channel Type. Like all cable trays, the channel type is prefabricated. It consists of a one-piece ventilated or solid-bottom channel section not exceeding 4 inches in width.

Solid-Bottom Type. This is a metal structure that has no openings in the bottom. The cable support is provided by integral or separate longitudinal side rails.

AERIAL INSTALLATION

The installation of aerial cable, or overhead as it is sometimes called, offers several advantages, including higher ampacity and low maintenance requirements. Several different types of installations are employed, including insulated triplex and quadraplex cables, noninsulated conductors on insulators, and solid- or pipe-type bus. The higher ampacity is a result of the excellent cooling properties of free air.

Aerial cables must be able to withstand the forces of high winds, storms, and ice loading. Self-supporting aerial cables may be strung directly from pole to pole, while other types must be lashed to a high strength steel wire called a messenger wire.

Pole Installations

Overhead installations are often supported by wooden, epoxy glass, or steel utility pole structures. When installed in such a manner, the conductors are supported on insulators, as shown in Figure 3–7. Such a structure may be used to install insulated or noninsulated wire for overhead systems. Other types of pole construction may be used for dead-ends, angles, and corners.

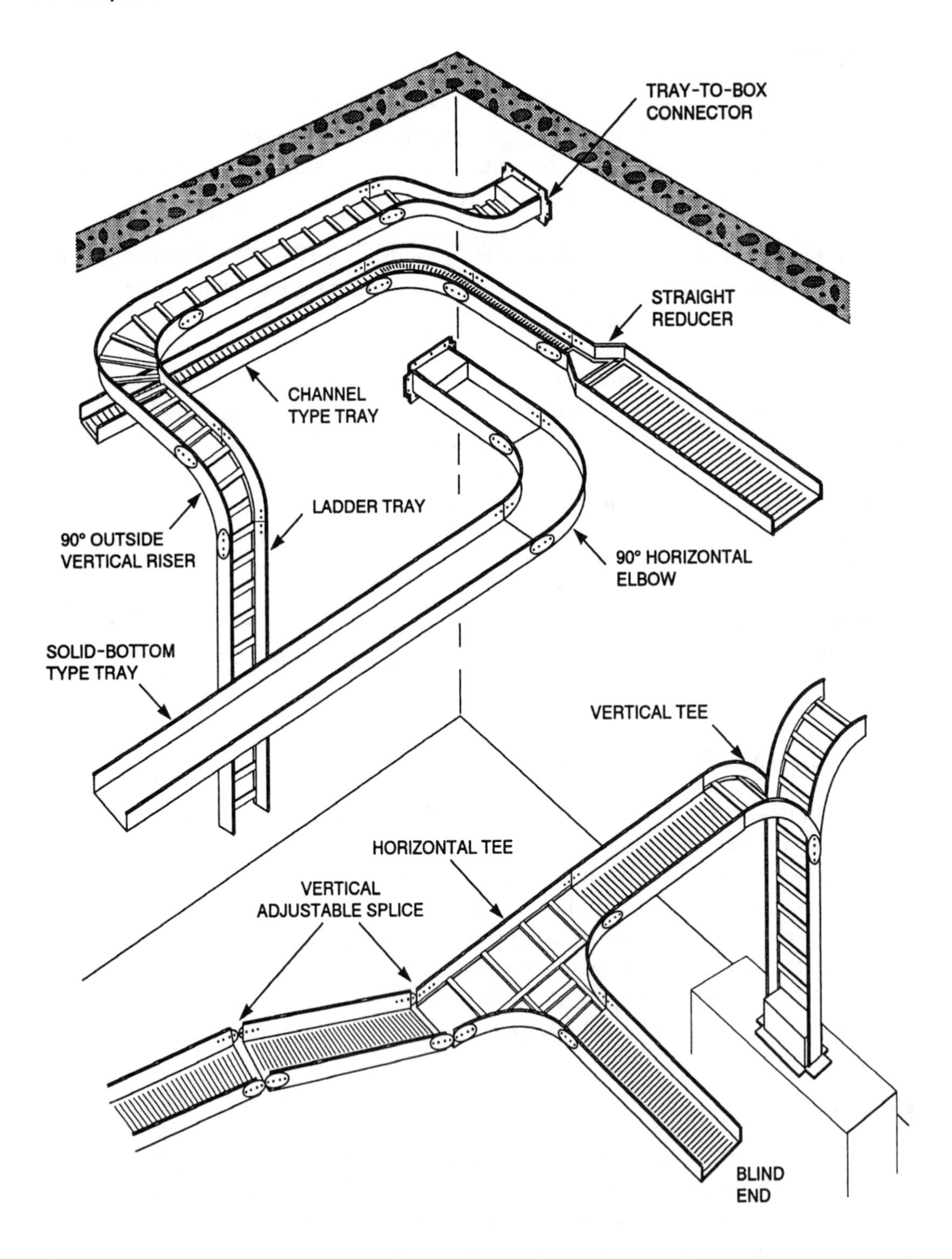

Figure 3–6 Cable tray system.

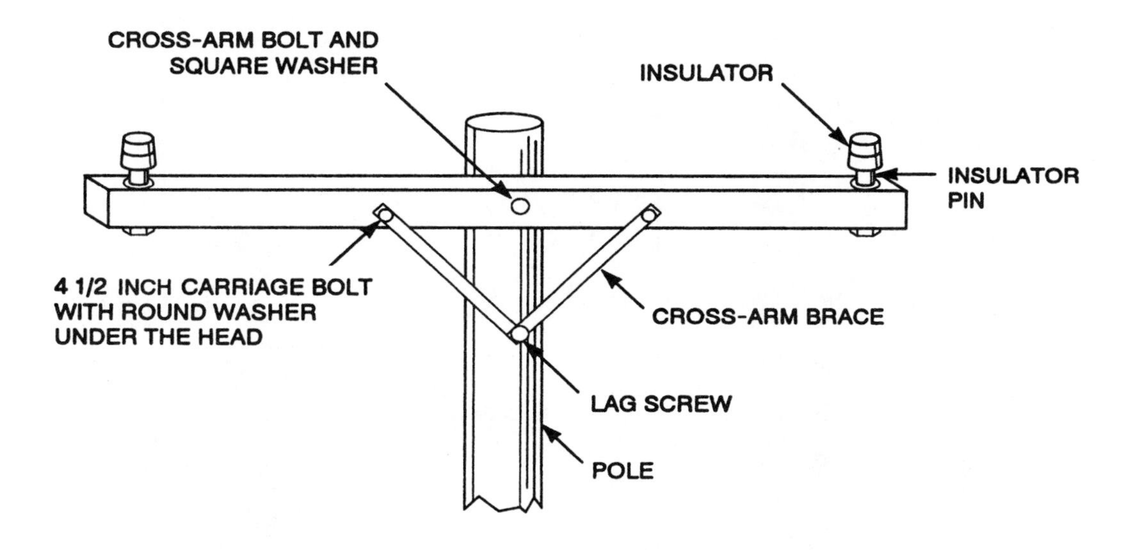

Figure 3–7 Type of cross-arm.

Some conductor spans are separated by insulating spacers. Such separators hold the conductors apart and stabilize them in high winds.

Triplex Cable

Triplex cable, as shown in Figure 3–8a, is a self-supporting cable that can be used for overhead runs. It is commonly used to connect utility distribution circuits to commercial or residential service entrances, as shown in Figure 3–8b.

DIRECT BURIAL INSTALLATION

Some cables are suitable for direct burial installation. Such cables are placed in the ground by plowing or trenching. A cable-laying plow opens the ground, lays the cable, and backfills the opening in one pass. In the trenching method, a backhoe or trencher is used to dig a trench into which the cable is laid. The trench is then backfilled. Figure 3–9 shows a cable-burying operation.

Buried cables must be protected against frost, water seepage, burrowing animals, and mechanical stresses caused by earth movement. Armored cables specially designed for burial are available. Cables should be buried at least 30 inches deep. This depth puts them deep enough to protect them from damage. In colder states, they are buried below the frost line. Other buried cables should be enclosed in sturdy polyurethane or PVC pipe. These pipes should have inside diameters several times the outside diameters of the cables to protect against earth movements. An

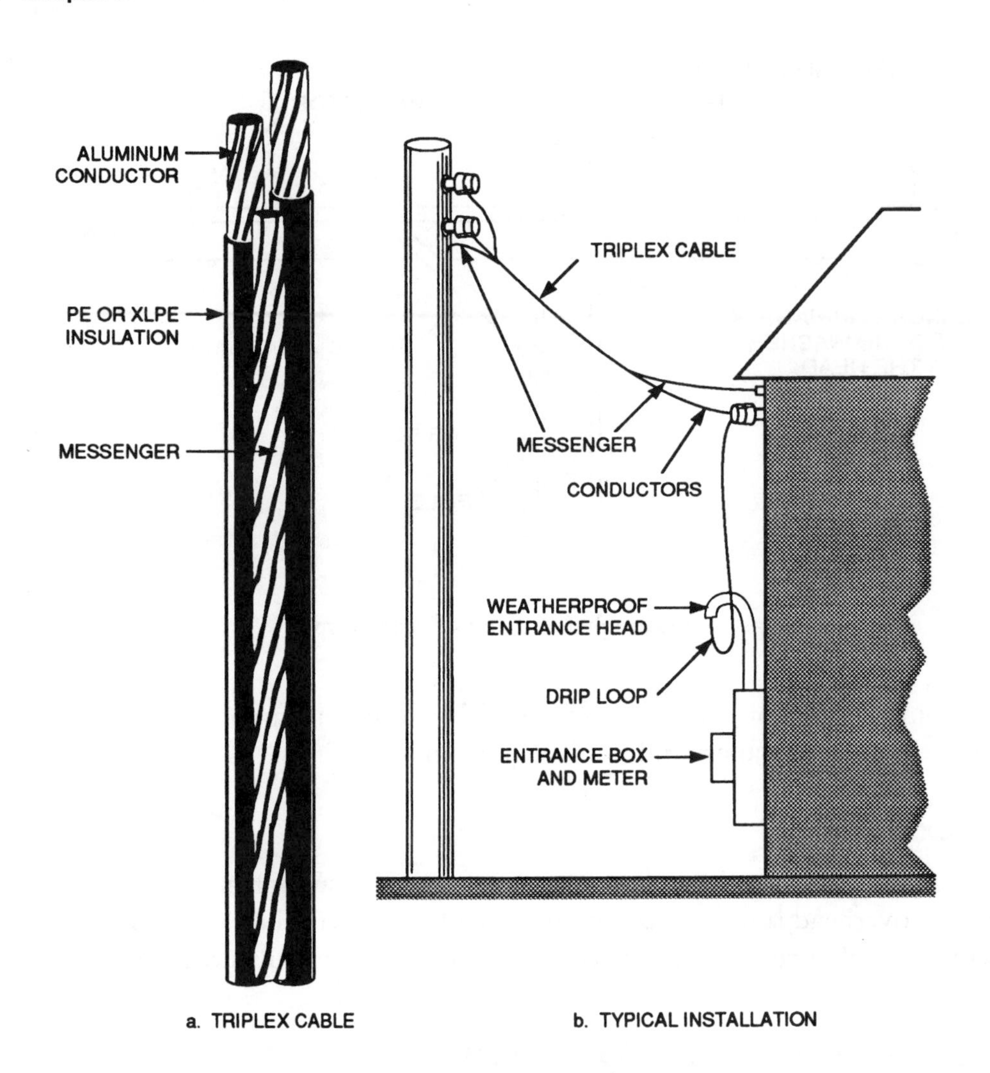

Figure 3–8 Triplex cable and installation.
(Courtesy of Cadick Corporation)

excess length of cable in the pipe prevents tensile loads from being placed on the cable as the system expands and contracts with temperature variations.

CABLE PULLING

Considerations

Cable may need to be pulled into the conduit, floor raceway, cable tray, wire mold, or metal raceway. Certain preparations must be made, regardless of the type of cable used or the application. At minimum, the cable puller should:

Figure 3–9 Burying underground cable.
(Courtesy of Alan Mark Franks)

- Check the *NEC*® and manufacturer's cable information charts to determine the suitability of the cable chosen for the job.
- Check the temperature rating and weight of the cable.
- Survey the environment and terrain.
- Measure the length of the cable run to determine the amount of cable required.

Insertion of Pull Line

Before a cable can be pulled into a raceway, a pull rope must be in place. Pull ropes are pulled into raceways by pull strings. The pull string is routed through the raceway using one of three common methods.

Existing Rope. When a cable is pulled into a raceway, a pull rope may be connected to it. The rope is left in place to pull other cables or pull ropes. This method requires care to avoid tangling the rope with the cable as it is pulled in. It is not suitable for small raceways.

Fish Tape. Fish tape has long been used to pull rope or string. The tape is "fished" into the raceway and connected to the pull string at the far end. The fish tape is then pulled back with the string connected.

Pressure/Vacuum. Figure 3–10 shows one of the best ways to insert the pull line. The use of a blower or a vacuum requires that the raceway be completely sealed. A lubricated plug is placed in the conduit with the pull string attached.

If a vacuum is being used, the hose is connected to the receiving end of the conduit and the plug/pull wire combination is literally sucked through the pipe. In the case of a blower, the hose is connected to the sending end and the plug/pull wire is blown through the conduit. Note that the blower type requires a special fitting on the hose end to allow passage of the pull wire.

Rigging

Regardless of whether cable is pulled by hand or machine, it must be properly attached to the pull rope. The method used will depend on the cable size, the length of the pull, the type of wire, manufacturer's recommendations, and other variables. The following sections describe the most common methods used for rigging the cable to the pull rope.

Taped Connections. Short, relatively lightweight pulls may be executed using a fish tape secured to the wire with electrical tape, as shown in Figure 3–11a.

Wire Mesh Grips. For more difficult pulls, a woven metal pulling grip (Figure 3–11b) may be used. The grip is pushed onto the insulation, with the eye connected to the fish tape or pull rope. The grip mesh is wound in such a way that pulling on it causes it to tighten on the cable. The harder the pull, the tighter the grip.

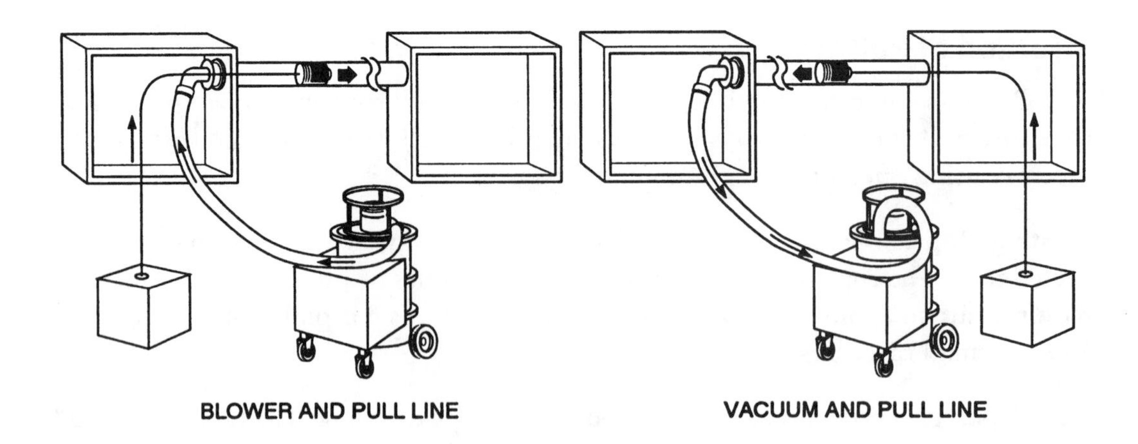

Figure 3–10 Using pressure or vacuum to insert pull line.

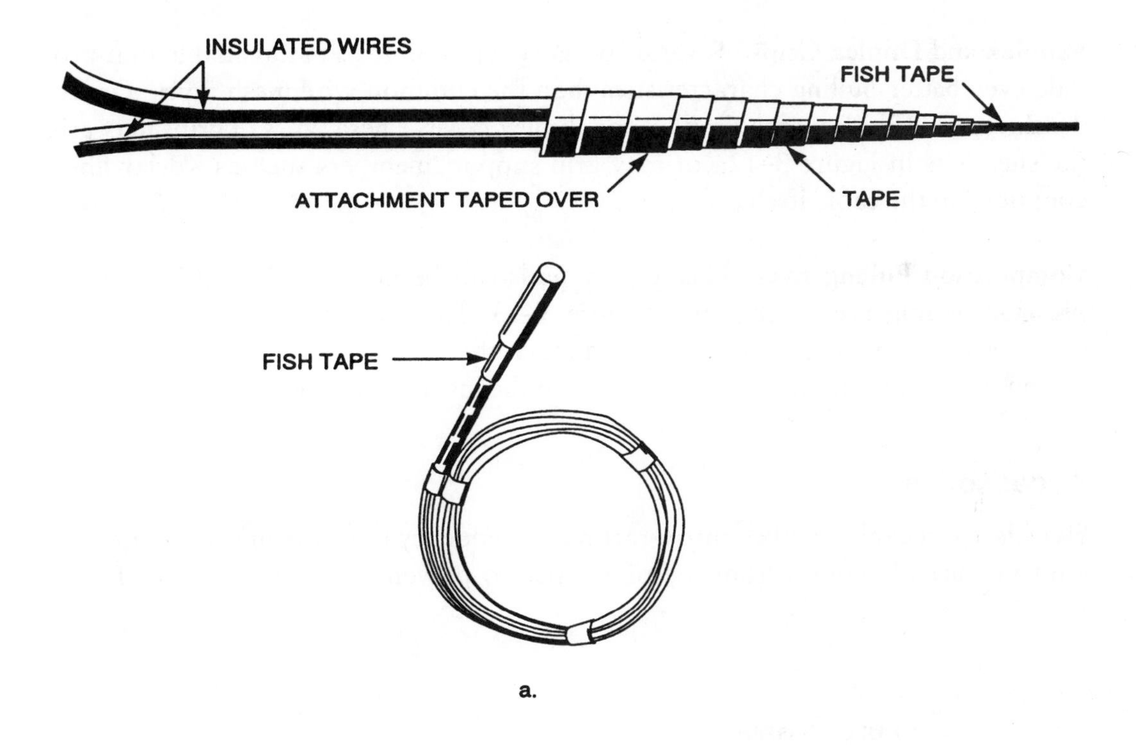

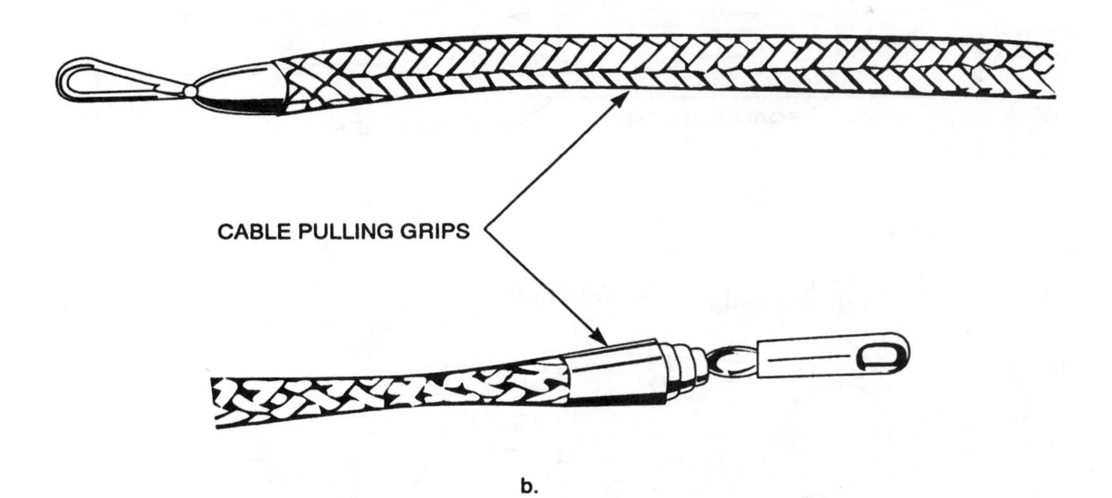

Figure 3–11 Wire pulling methods: a. fish tape and electrical tape connections and b. wire mesh grips.

Simplex and Duplex Grips. Several specialty types of grips are available that provide even better pulling characteristics than the common wire mesh types. Figures 3–12a and 3–12b show the simplex and duplex types. These types of grips may grip the sheath as in Figure 3–12a or may grip support members such as Kevlar fibers contained in the cable itself.

Compression Pulling Eyes. Heavy-duty pulls can be performed by using a compression pulling eye, as shown in Figure 3–13. This is a special assembly that is compressed onto the cable conductor in much the same way as an electrical termination lug. Some pulling eyes compress onto the insulation as well as the conductor.

Accessories

Swivels. As a cable is pulled into a raceway, a tendency to rotate may develop. This tendency usually comes from the pull rope. To prevent this, a swivel (see Figure

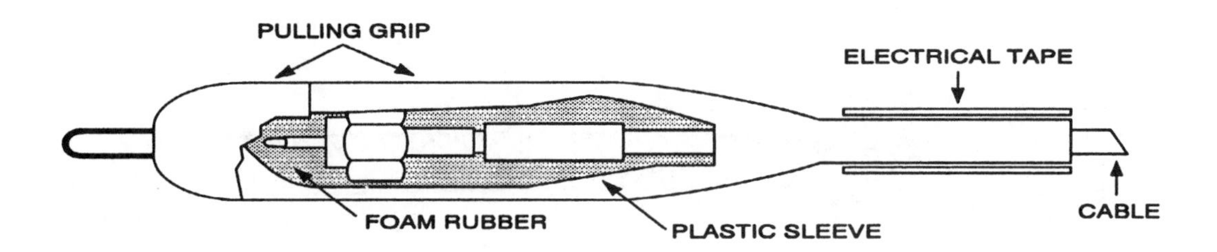

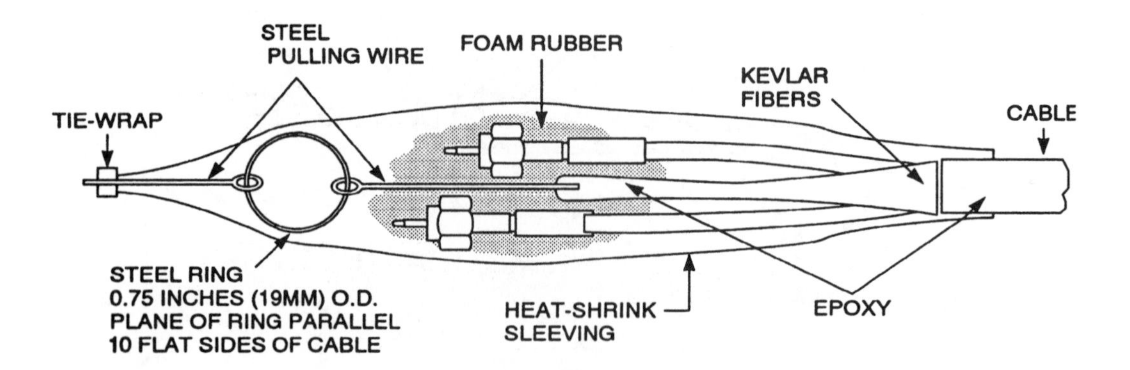

Figure 3–12 Pulling grips: a. simplex pulling grip and b. duplex pulling grip.
(Courtesy of Canoga-Perkins Corporation)

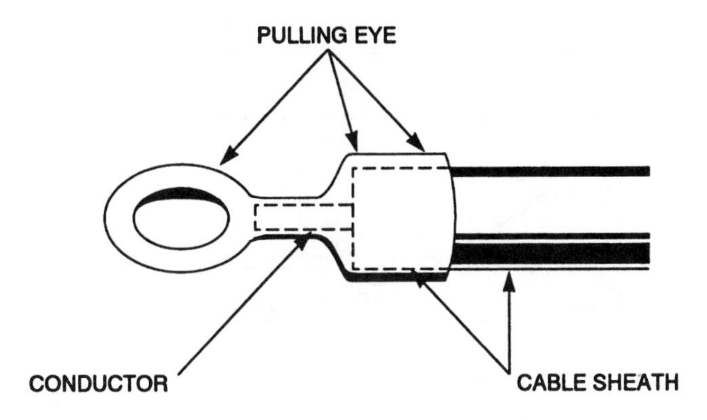

Figure 3–13 Compression pulling eye. *(Courtesy of Cadick Corporation)*

3–14) is often used, which is free to rotate. Thus, as the pull rope turns, the swivel rotates and prevents the cable from being twisted.

Some swivels are equipped with a shear pin assembly, which will break if the pulling tension exceeds a certain value. Many pulls require the use of this feature to prevent excessive and damaging pull tension.

Sheaves. Sheaves are pulley assemblies that are used to change the direction of the pull force. They are particularly useful for pulling cable into underground ducts that terminate in manholes or other such structures.

Many manholes are equipped with eye bolts, which can be used to mount simple pulley mechanisms. Others are not so equipped, and a guide-sheave rack must be used. Figure 3–15a shows a typical guide-sheave rack, and Figure 3–15b shows an installation of a rack.

When using sheaves, the following factors should be considered:

- The sheaves and the rack must be the proper size for the cable that is being pulled. A sheave that is too small will cause the cable to ride on the pulley ridges and damage the cable. A sheave that is too large will cause the cable to move back and forth and cause an unsteady pull.

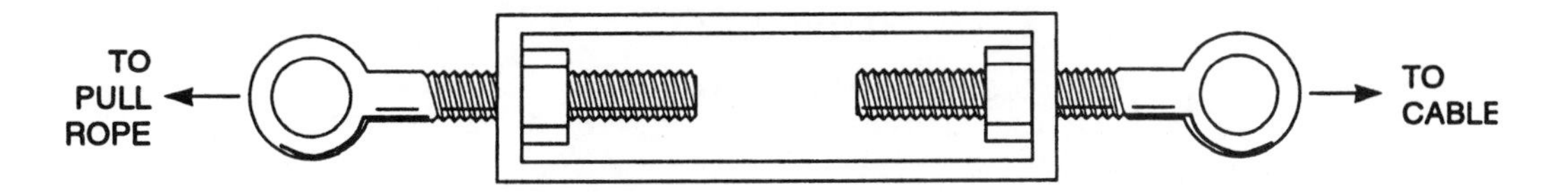

Figure 3–14 Swivel.
(Courtesy of Cadick Corporation)

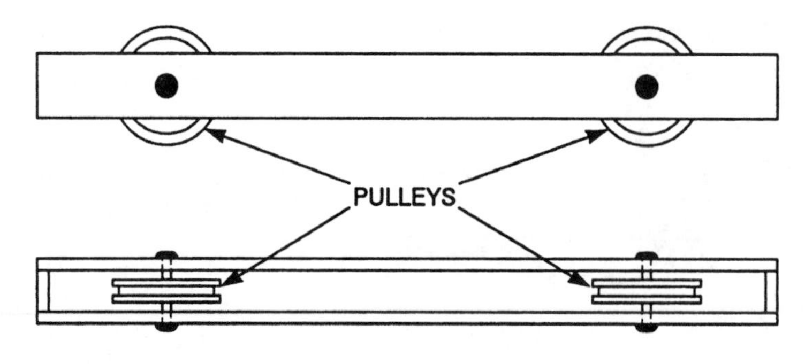

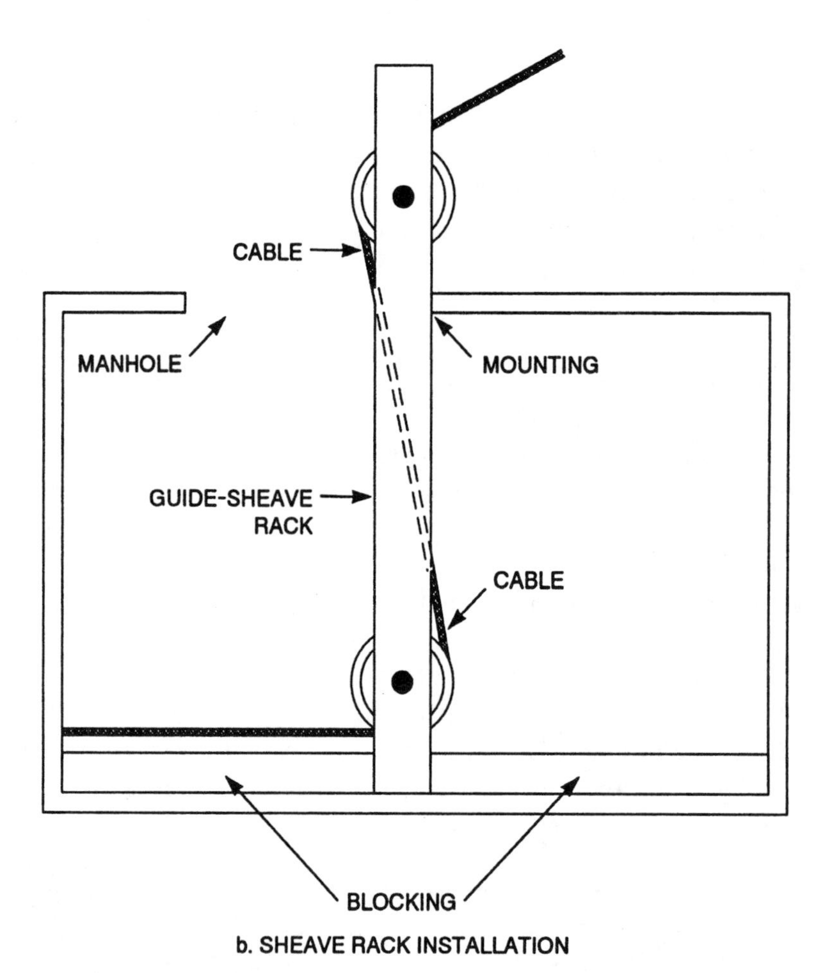

Figure 3–15 Typical guide-sheave rack and installation.
(Courtesy of Cadick Corporation)

- Sheaves increase the sidewall pressure on the cable during the pull. Refer to the manufacturer's pull tension instructions when using sheaves.
- Be certain to adequately secure the guide-sheave rack. If the rack is inadequately secured, it can slip, causing cable damage or personal injury.

Protectors. Protectors are metal or leather fittings that are placed into the lips of raceways. As the cable is pulled into the raceway, it rides on the protector and is protected from damage. Manufactured metal protectors may be purchased, but many electricians choose to fabricate leather ones in the field.

Pulling Machines

On longer, more difficult runs, cable-pulling machines may be employed. Figure 3–16 is a hand-operated pulling machine. This machine operates somewhat like a fishing reel. The cable is connected to the metal cable and then pulled through the conduit by turning the crank on the pulling machine.

For extremely difficult pulls, the motorized cable-pulling machine shown in Figure 3–17 may be used. The motorized machine (see Figure 3–18) is started by mounting it onto a stable pipe or stud using the anchoring chains. The rope line is then wrapped around the capstan, and the machine is started.

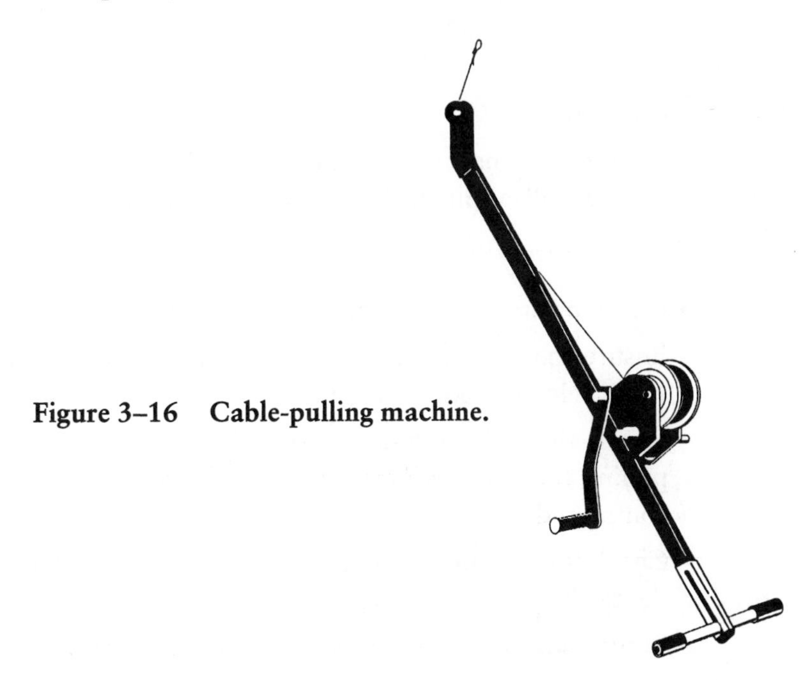

Figure 3–16 Cable-pulling machine.

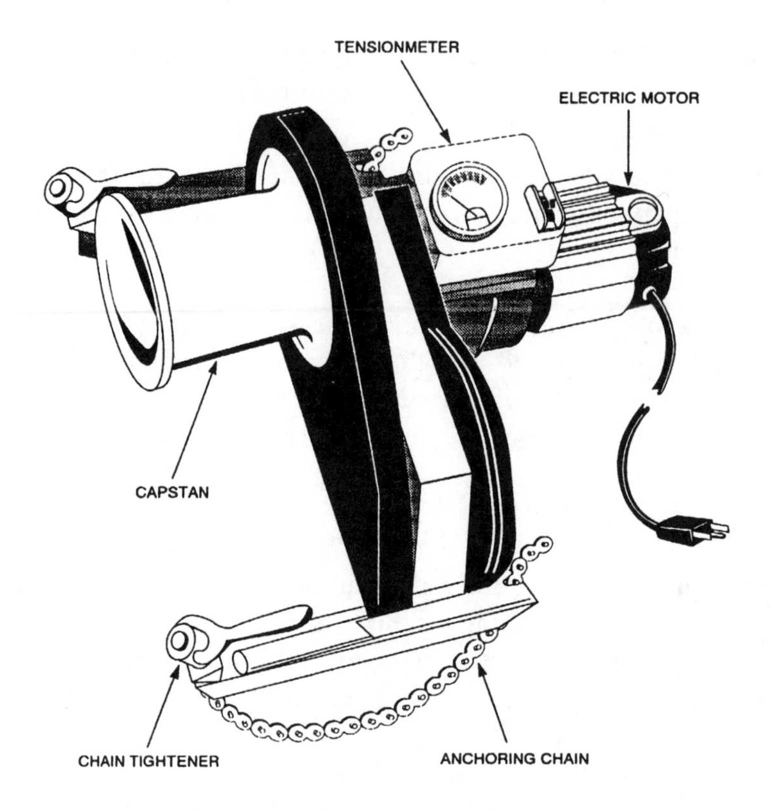

Figure 3–17 Power cable puller.

Cable Preparation

Always refer to the manufacturer's specifications for the particular type of cable and cable pull applied. The following general practices should always be observed:

- Check prints to determine the proper routing for the cable.
- Ensure that the splice and terminating prints are correct for the cable being pulled.
- Check that all necessary materials are available, such as cable pulling tools and equipment, rope, knife, and proper lubricating compounds.
- Check that the pull will occur at a temperature that is suitable for the cable type.
- Verify that the proper color codes are being used.

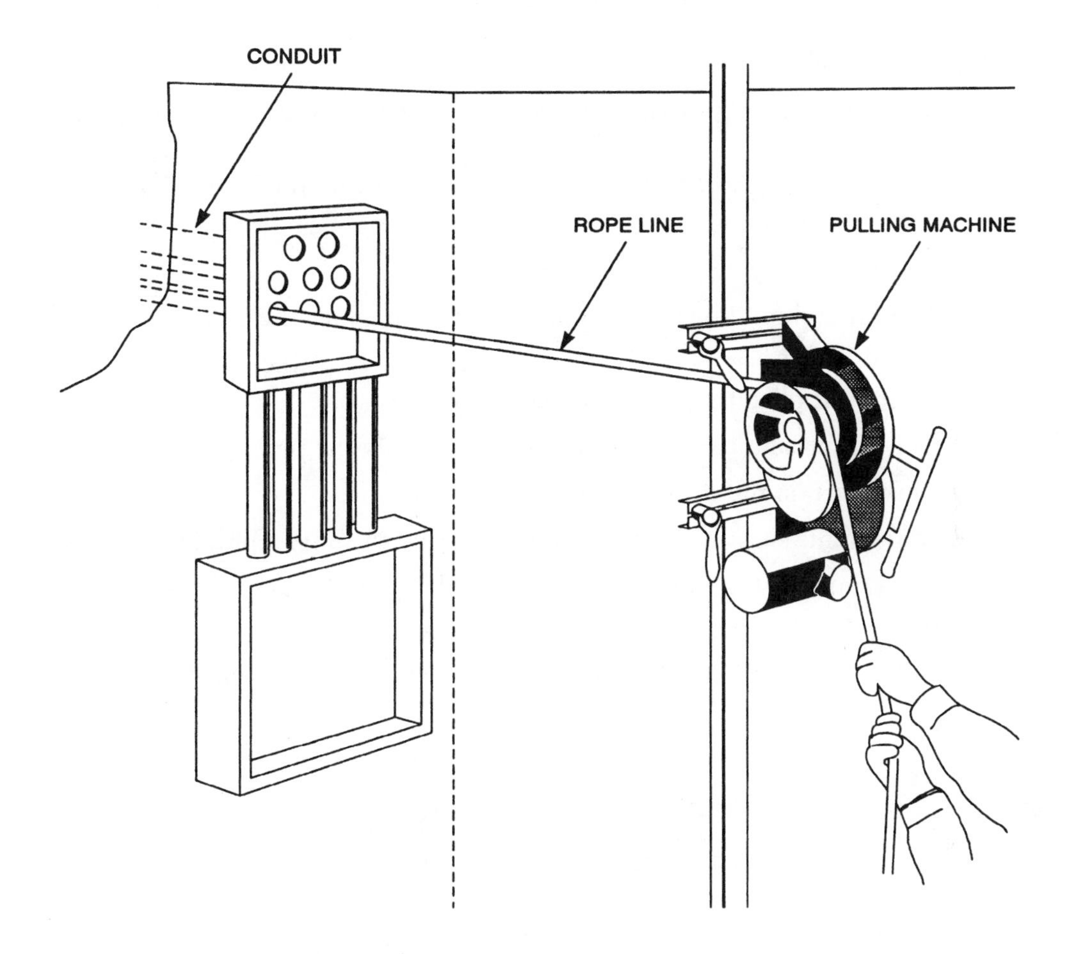

Figure 3–18 Cable-pulling machine in use.

- Determine the correct pulling direction.
- Calculate the cable sidewall pressure for each pull. Manual calculations or computer programs may be used for this step. Refer to the cable manufacturer or specialty manufacturer literature for details.
- Position the cable reels properly and watch for any sharp bends or kinks in the cable. Use reel jacks and axles to minimize friction when pulling and feeding the cable.
- Determine the proper number of personnel required for feeding the cable into the conduit.
- Be sure an ample supply of proper lubricant is available.

Pulling the Cable

The following general practices should always be observed when pulling cable:

- Two-way voice or hand-signal communications must be established prior to pulling.
- Short runs may be pulled by hand.
- The cable should be fed as straight as possible into the conduit.
- If basket grips are to be used (see Figure 3–11b), sufficient tension must be applied to get a positive grip on the cable.
- Apply lubricant to the cable constantly during the pull.
- Make certain the sheaves are the proper size.
- Leave some slack if the cable is to be pulled through a pull box.
- Be certain to tag and identify the cable at pull boxes, junction boxes, and terminal points.
- Correct pulling tensions must be calculated *before* pulling starts.

Cable-Pulling Precautions

The following general practices and cautions should be observed:

- During the pull, avoid cable cross-over where the cables enter the conduit.
- Observe special care when pulling cables through manholes where damage may occur. Use guide-sheave racks where required.
- Do not allow vertical conduit rises to be supported by conduit fittings (condulets).
- Do not hammer cable into a fitting. Such hammering can severely damage the cable.
- Consider sidewall pressure for vertical hanging cables bent around a corner.
- The direction of the pull and the number of bends affect the tension and sidewall pressure.
- Be certain that the conduit or cable tray is cleared of debris and foreign particles.

- On long pulls, pull from a pull box in the center. This reduces the pull tension by half.
- Use a protector inserted in the conduit to avoid abrasion as the cable is pulled in.
- When pulling cable into enclosures with energized conductors, use barriers and guards to be certain that inadvertent contact is not made. Observe Occupational Safety and Health Administration (OSHA) "Safety Related Work Practices" clearance distances.

Installing Cable in Cable Tray

Cable installed in trays should be pulled over rollers that are spaced according to the weight of the cable being pulled, as shown in Figures 3–19 and 3–20. These rollers relieve friction and equalize the tension on the cable as it moves through the tray. The following practices should also be observed:

- Use sheaves and cable grips to make the pull.
- Use cover protection during construction.
- Use tension-limiting tools if nylon support grips are applied to the cable.

Tie-Offs and Supports

Supports should be provided to prevent mechanical stress on cables where they enter another raceway or enclosure from cable tray systems.

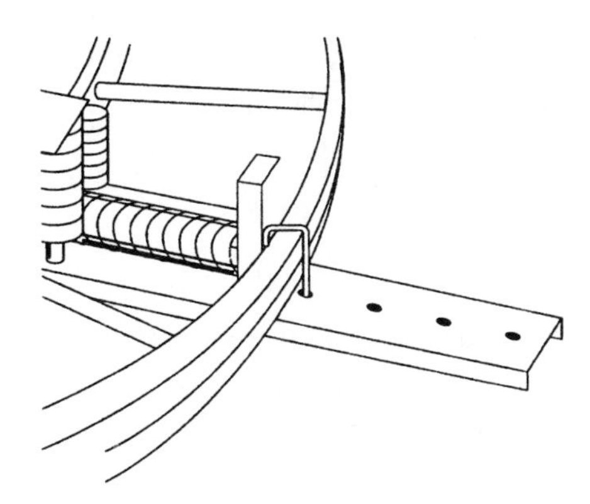

Figure 3–19 Cable-pulling rollers.

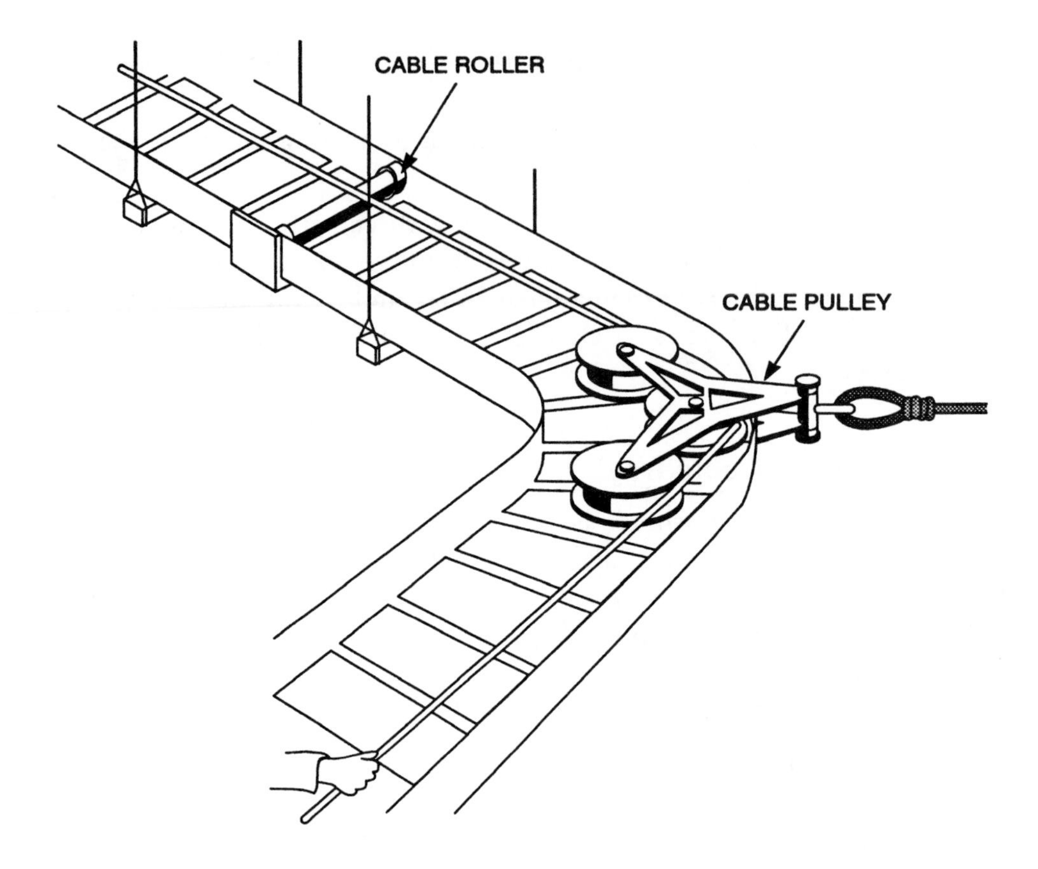

Figure 3–20 Pulling cable in the cable tray.

In other than horizontal runs, cables should be fastened or tied off securely to transverse cable trays.

Single conductors should be bound in circuit groups to prevent excessive movement due to fault current magnetic forces. Some manufacturers require that cable be tied off or laced down before equipment will meet Underwriters Laboratories® (UL®) standards.

Lubricants

Lubricants must be used to reduce the friction of the cable as it moves through the raceway. Many lubricants are used in the field, including soap, motor oil, and grease. The following sections discuss four of the most commonly used commercially available lubricants.

Wax/Glycerin/Water Emulsions. These lubricants provide good friction reduction but leave messy residues. After the residue dries, it may cause the cable to stick in the conduit, making removal difficult or impossible. Some of the wax lubricants can cause cracking of some cable jackets.

Bentonite Clay/Water. These lubricants are slurries of naturally occurring clay in water with or without the addition of ethylene glycol. While they are low in cost and do not damage the cable insulation, they do not provide as much friction reduction as other lubricants.

Polymer/Water. Polymer/water lubricants are moderately priced lubricants that are easy to clean. Their unique molecular structure creates an elastic consistency that allows the lubricant to stay with the cable as it pulls through the conduit. They have very low friction and leave little residue.

Hydrocarbon Grease. These lubricants are greasy, jelly-like substances that are obtained from petroleum. While very low in friction, they are quite expensive and extremely difficult to clean from clothing, hands, and cable.

Pulling Tensions

Calculating maximum and actual pulling tensions must be performed on an individual basis. Always refer to the manufacturer's literature for specific information. In general, the following apply.

Maximum Allowed Tension. The following directions, supplied courtesy of General Cable Company, may be used as a reference for maximum allowed pulling tension.

- The maximum tension should not exceed 0.008 times the circular mil (CM) area when pulled with a pulling eye attached to copper or aluminum conductors. The formula used is:

$$T = 0.008 \times n \times CM$$

Where:

T = maximum tension in pounds
n = number of conductors in cable
CM = outside diameter of cable in inches

- The maximum stress for lead cables shall not exceed 1,500 pounds/square inch of lead sheath area when pulled with a basket grip.
- Maximum tension shall not exceed 1,000 pounds for nonleaded cables pulled with a basket grip, but the maximum stress calculated from the preceding formula cannot be exceeded. When pulling a one-conductor (1/C) cable, maximum tension must not exceed 5,000 pounds. Maximum tension for two or more conductors shall not exceed 6,000 pounds.
- Maximum tension at a bend must not exceed 300 times the radius of curvature of the duct expressed in fee, while also not exceeding maximum tension calculated from the first three criteria. Therefore, the minimum radius should not be less than:

$$R(f) = \frac{T}{300}$$

Where:

$R(f)$ = conduit radius of curvature
T = maximum pulling tension from the first criteria or manufacturer's data

Note that various types of cables have different bending radii. See Chapter 6 for detailed information about each cable type.

Calculating Tension. Pulling tension in a horizontal duct may be calculated from the formula:

$$T = L \times w \times f$$

Where:

T = total pulling tension
w = weight of the cable in pounds/foot
f = coefficient of friction

See the manufacturer's literature or cable-pulling handbooks for calculating tension in curved ducts and more complicated structures.

EXPOSED CABLE INSTALLATION

Exposed cable is vulnerable to environmental or physical hazards because it is not protected by metal raceway, conduit, or other enclosures. The *NEC*® requires that conductors be adequately protected where subject to physical damage. The types of protection needed are as follows:

- Protection against penetration by screws or nails by the use of a steel plate or bushing of the appropriate length and width.
- When the cable has to pass through wood members, holes must be drilled or notches made.
- When cables pass through metal framing members and behind panels, they should be protected by bushings or grommets securely fastened in the opening.
- Cables routed underground or beneath buildings must be in a raceway that extends beyond the walls of the building.
- Any cables or conductors that emerge from the ground must be protected by an enclosure or a raceway.
- Conductors or cables that enter a building must be protected at the point of entrance.
- Cables must also be protected if they are in the immediate area of any activities where people are walking with tools or equipment.

Cables must also be properly secured and supported. Types of supports include clamping devices, tie-down straps, cable metal, plastic straps, and cable hangers. Supports must be properly spaced and mounted depending on the size and length of cable run.

Raceways, cable assemblies, boxes, cabinets, and fittings must also be securely fastened in place.

WIRING INSTALLATIONS IN HAZARDOUS LOCATIONS

A hazardous location is any location in which a potential fire or explosion hazard may exist because of the presence of flammable, combustible, or ignitable materials.

The *NEC*® classifies hazardous locations according to the properties and quantities of the hazardous material that may be present. Hazardous locations are divided into three classes (I, II, and III), two divisions (1 and 2), four classified groups (A, B, C, and D), and two unclassified groups (E and G). Any installations in these areas must be suitable for the locations. Mineral-insulated cable (Type MI) is the only cable permitted for use in *all* hazardous locations. Other cable types may be used in one or more of the various locations.

Refer to the *NEC*® for detailed information.

INSTALLING COMMUNICATIONS CABLES

The requirements for installing communications cables are essentially the same as those for power cables outlined earlier in this chapter. The following key points identify the significant differences between the two types of cables. More detailed coverage of this material may be found in *Premises Cabling* by Donald J. Sterling, Jr.

Minimum Bending Radius

Figure 3–21 illustrates the bending radius. Regardless of the type of cable, the bending radius must be large enough to prevent damage to the cable or its insulation. This is true both during the installation of the cable and during normal usage with vibration, chemicals, and other environmental conditions.

The bending radius must be above a given value in communications cable for another important reason. Copper communications cable with too sharp a bend can change characteristic impedance, thereby changing operating characteristics and possibly causing attenuation to rise dramatically.

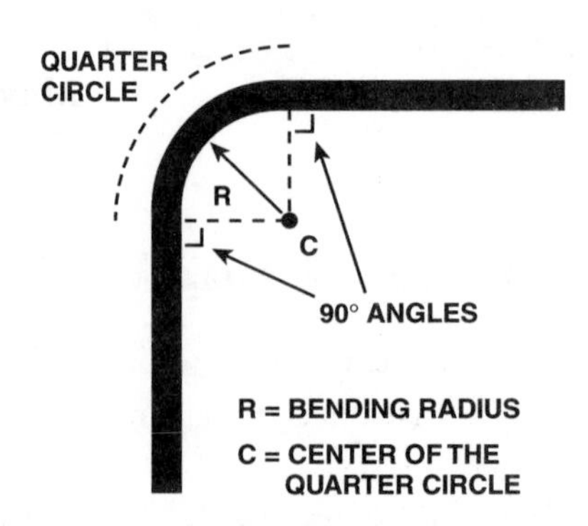

Figure 3–21 Definition of the bending radius.
(Courtesy of Cadick Corporation)

Fiber cable that is bent too sharply reflects light improperly. That is, the light beam strikes the interface at too great an angle and is not reflected, which greatly increases attenuation.

Table 3–1 identifies the minimum bending radii for common communications cable types. Note that only values for Category 5 are used because it has the most critical bending radius requirements.

Table 3–1 Bending radii for communications cables.

Cable Type	Bending Radius (number of times the cable diameter)	Comments
UTP (4-pair horizontal)	4	1 inch minimum (TIA/EIA–568A)
UTP (multiple pair)	10	
Optical fiber	4	About 1.8 inches for zipcord construction

Pulling Tension

The maximum pulling tension for communications cable is related primarily to physical damage. Excessive tension damages all types of cable and causes them to fail. Communications cable that is pulled with too much tension can degrade in performance without actually failing. For example, Category 5 cable that is pulled improperly can stretch and degrade to Category 4 or lower performance characteristics (see Table 3–2).

Table 3–2 Pulling tension for communications cables.

Conductor	Note	Maximum Loading per Conductor (in Pounds)
#22 AWG Copper	Solid	2.0
	Stranded	2.2
#24 AWG Copper	Solid	3.2
	Stranded	3.6
#26 AWG Copper	Solid	5.0
	Stranded	5.6
Optical fiber	Installation	100 (typical) to 210
	In service	14 (typical) to 65

Other Points to Consider

- Avoid deforming cable with mounting hardware.
- Keep communications cable away from electromagnetic interference sources. Table 3–3 shows TIA/EIA–569 recommendations for electrical power lines above 480 volts.
- Keep conduit fills to 50 percent or lower.
- Avoid splices.

Table 3–3 Recommended separation from high-voltage power sources.

Condition		Minimum Separation Distance (in Inches)		
Power Line Pathway or Electrical Equipment	**Signal Line Pathway**	**Less than 2 kVA**	**2–5 kVA**	**Over 5 kVA**
Unshielded	Unshielded	5.0	12	24
Unshielded	Shielded	2.5	6	12
Shielded	Shielded	—	3	6

Shielded: Metallic, grounded conduit
Unshielded: Open or nonmetallic conduit

SPLICING

Overview

Weight and size limitations dictate the amount of cable that may be placed on a reel. Because of this, most installations require that cable be spliced so that runs are long enough. The key feature of a properly made splice is that the splice has equal or better mechanical and electrical characteristics as the cable being spliced. This section illustrates several common types of splices.

Although low- and medium-voltage splices generally have the same electrical requirements, the shield on medium-voltage cable must be taken into consideration. Because of this, low- and medium-voltage splices are discussed in separate sections.

Low-Voltage and Nonshielded Splicing

Screw-On Pigtail Connectors. Screw-on pigtail connectors, also called wire nuts, (see Figure 3–22), are among the most popular and convenient ways to splice wires.

Wire nuts are threaded nuts with plastic insulating covers. To use them, the two or three wires to be spliced are first stripped to a distance of approximately ½ inch and then placed parallel to each other. The wire nut is threaded onto the three wires and tightened in much the same way as is a wing nut. The resulting connection is both mechanically and electrically sound.

Wire nuts are used in commercial applications to join fixture wires and branch circuit wires to fixture wires. If the wire nut is not equipped with a metallic coil spring, it may not be used for general-purpose branch-circuit wiring.

Bolted Pressure. In some lower voltage power applications, bolted pressure connector sleeves are used. Figure 3–23 identifies four basic types of such connectors.

Figure 3–22 Wire connectors.

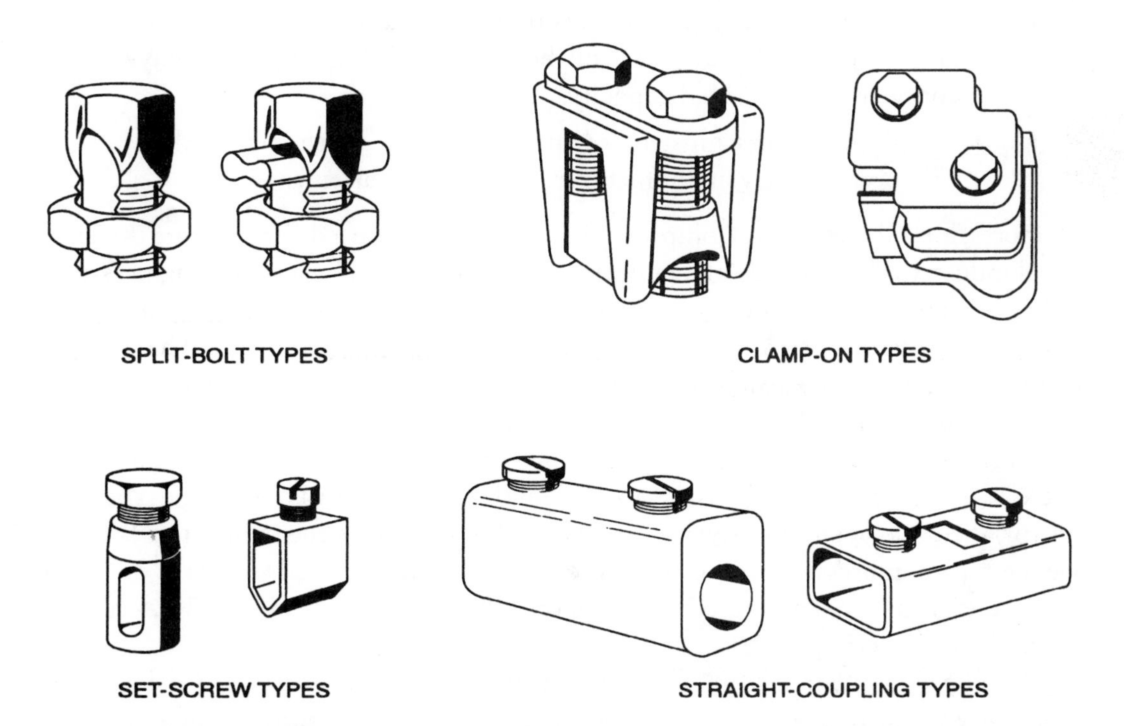

Figure 3–23 Types of mechanical pressure splice connectors.

Bolted pressure connectors depend on the applied force of bolts or screws to produce the necessary clamping and contact pressure between the conductors and the connector.

For some connectors, the only tool required to compress the conductor is a screwdriver. Connectors can be reused and are recommended for use in installations in which wiring connections will be changed frequently. Various set screw connectors with thread-on insulated caps are suitable as pressure wire connectors or fixture-splicing connectors, according to individual manufacturers' listings.

The material from which connectors are made varies depending on the type of wire to be connected. Copper, bronze, and copper/bronze alloys are often used for connectors of copper wire. Aluminum, tin-plated silicon-bronze, or tin-plated copper alloys are used for aluminum wire. Bolted-type connections have a few disadvantages, including that:

- They are subject to the mechanical failure of the screws.
- Bolts may not adequately engage the wire strands.
- Connectors with copper bodies or steel bolts may not give a reliable connection.
- ANSI/IEEE Standard 80 requires that some bolted connections be reduced in ampacity. This means that a bolted pressure connector may become a "weak link" in the chain and that the cable may not be used to its rated capacity.
- Bolted connectors must be torqued to proper levels before they will meet industry, manufacturer, and code requirements.

Compression Connectors. Compression connectors are used throughout the electrical industry. They include all connectors that are applied by a squeezing pressure, either by hand or by a pneumatic or hydraulic compressing device. Remember that many compression connectors require the use of a no-oxide lubricant. This is especially important when using aluminum conductors. Refer to the manufacturer's instructions.

Figure 3–24 shows various types of low-voltage, compression-type splice connectors. To use them, a properly stripped length of conductor is inserted in the connector, then a proper compression tool is used to compress the connector. When a wedge-type compression tool is used, the wedge must be placed so that it does not crack the seam on the connector. The connectors shown in Figure 3–24 are all examples of the crimp-on type of compression connector.

Compression connectors as shown in Figure 3–25 can be used for low- and medium-voltage applications. Such a connector is shown properly applied in Figure 3–26.

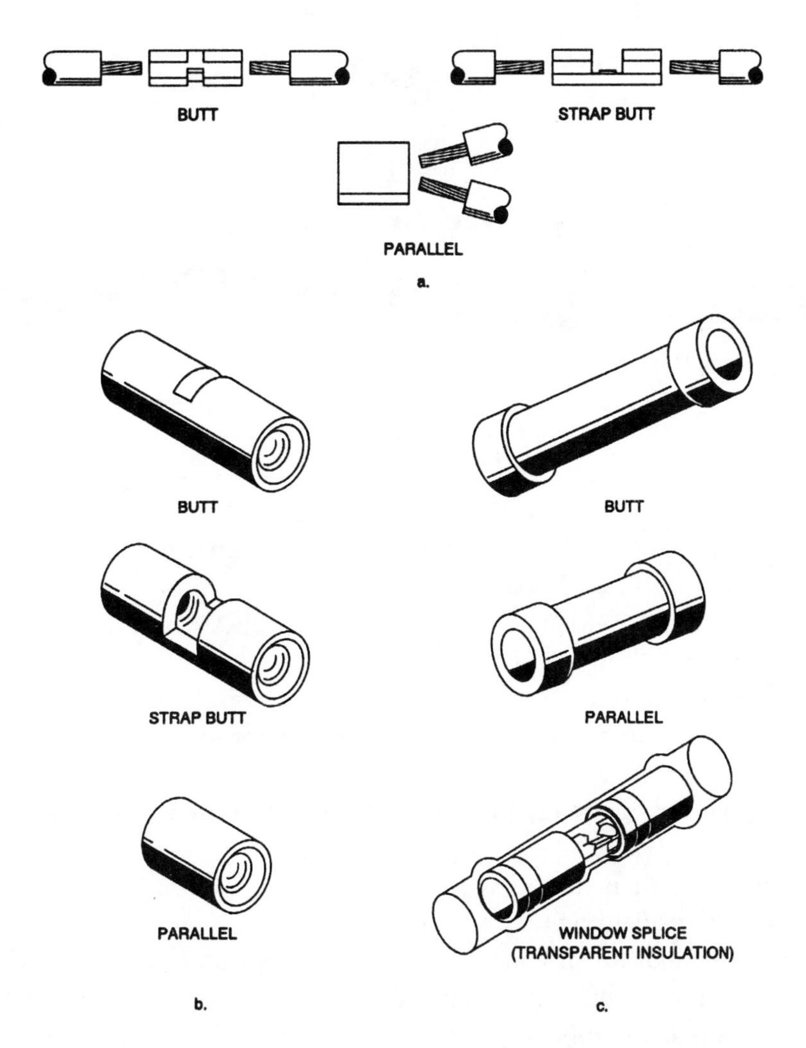

Figure 3–24 Low-voltage, crimp-on splice connectors: a. crimp-splice types, b. noninsulated crimp-on splice connectors, c. insulated crimp-on splice connectors.

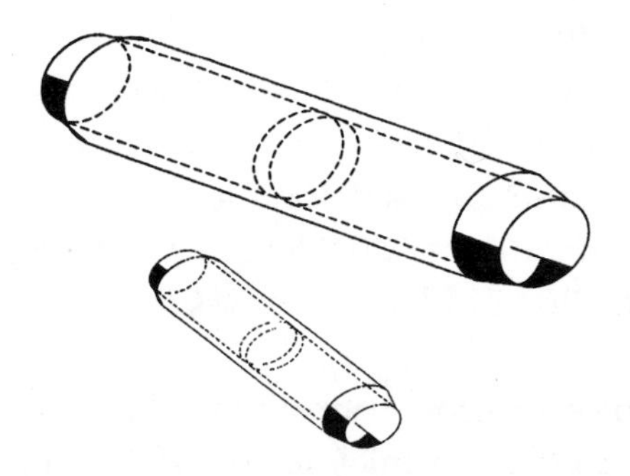

Figure 3–25 Compression connectors.

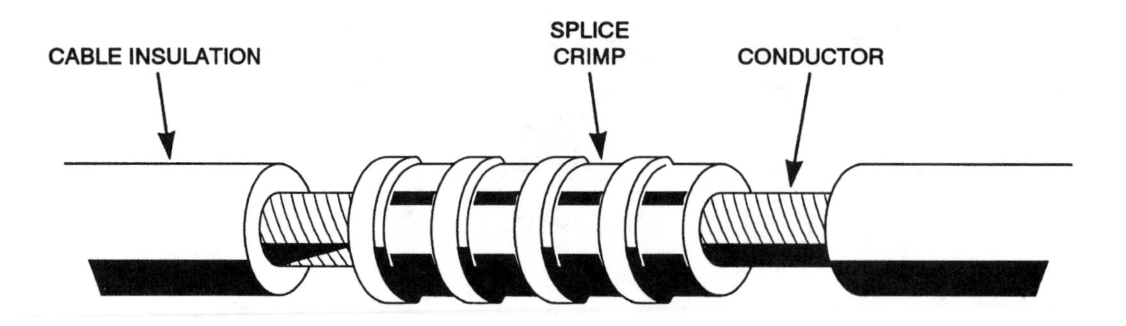

Figure 3–26 Compressed connector sleeve.

Splicing Procedure. The most difficult part of any splice is the completion of the insulating system. Figure 3–27 shows the correct procedure for a 0–5kV cable. This procedure applies to PE, XLP, and EPR insulation. The following procedure refers to Figure 3–27 and is provided courtesy of Rome Cable Corporation.

1. Train conductors into position. Cut ends squarely at centerline of splice.
2. Measure the distance *B* + *C* + *D* + ½ *connector length,* and mark each conductor.
3. Remove jacket and underlying cable tape (if present) from each end of the cable.
4. Remove cable insulation for a distance *B* + ½ *connector length.* Remove conductor shielding tape (if present).
5. Pencil insulation of each end of distance C. Smooth with file or garnet cloth, making sure shielding tape (if present) is cut squarely at ends of pencilled insulation.
6. With a knife, file, or garnet cloth, round exposed edge of jacket to eliminate sharp corners.
7. Install connector as specified.
8. Clean jacket, insulation, and connector with a clean cloth that has been moistened with a nontoxic solvent such as trichlorethylene.
9. Fill connector indents with pieces of conducting shielding tape.
10. Cover connector and exposed conductor area with one half-lapped layer of conducting shielding tape. Note that the tape should overwrap the ends of the pencilled insulation for a distance of 1⁄16 inch *only* (dimension G).
11. Measure dimension *E* from each end of the jacket and pencil cable.
12. Apply insulating tape to a thickness *H* using uniform tension throughout, and apply in half-lapped layers between the pencils made in Step 11.

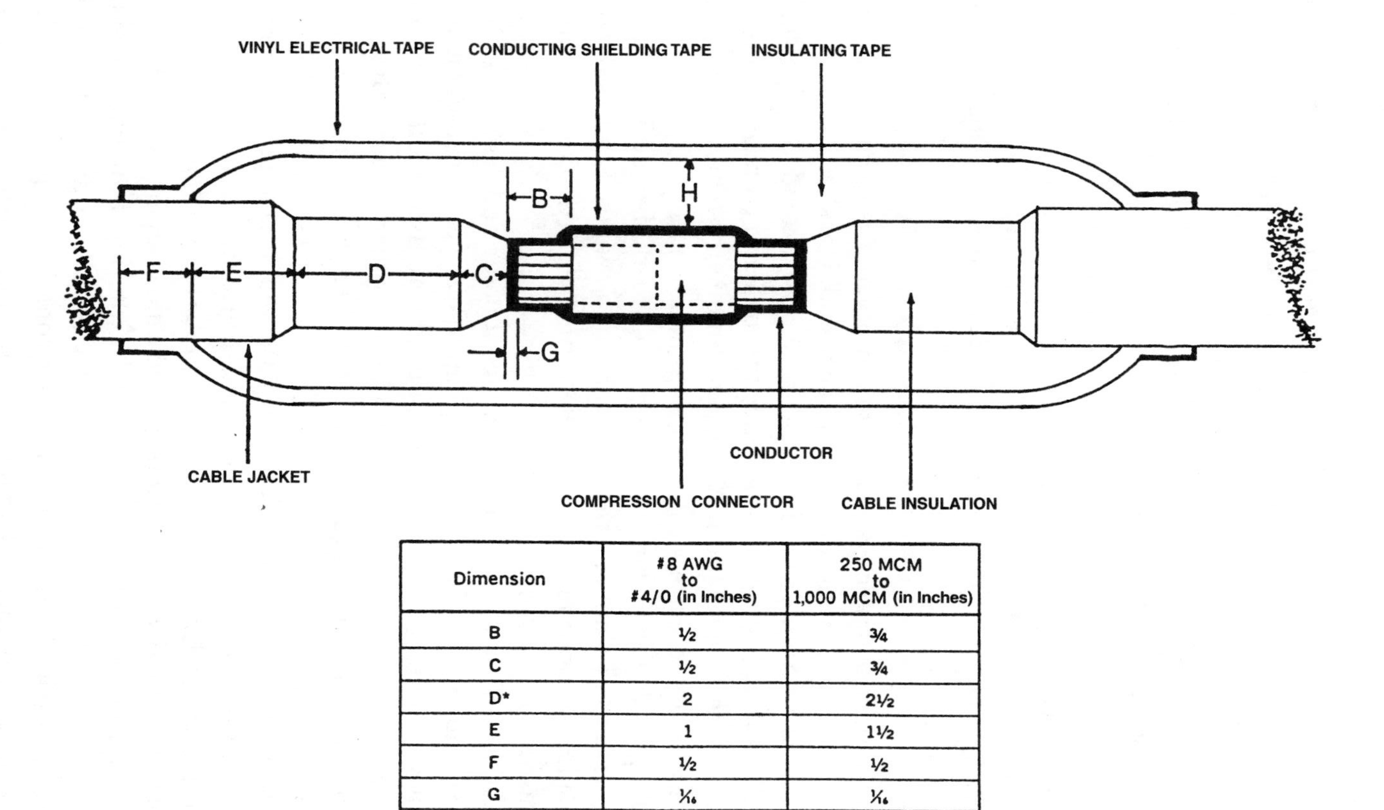

Dimension	#8 AWG to #4/0 (in Inches)	250 MCM to 1,000 MCM (in Inches)
B	1/2	3/4
C	1/2	3/4
D*	2	2 1/2
E	1	1 1/2
F	1/2	1/2
G	1/16	1/16
H	1/4	3/8

*D = D + E when there is no jacket.

Figure 3–27 Splicing 0–5 kV cable.
(Courtesy of Rome Cable Corporation)

While taping, stretch the tape to three-quarters its original width. The diameter of the taped areas over the connector should equal *(2 × H) + (connector diameter) + 1⁄16 inch.*

13. Cover the taped area with two half-lapped layers of vinyl electrical tape. The tape should extend ½ inch beyond the insulating tape at each end of the splice.

Medium-Voltage and Shielded Splicing

The actual joining of the conductors for this type of splice is very similar to low-voltage and nonshielded splicing. The actual procedure has five steps: (1) preparing, (2) joining conductors, (3) reinsulating, (4) reshielding, and (5) rejacketing. The following explanation of these steps is provided courtesy of the 3M Corporation.®

Preparation. Begin with good cable ends. A common practice is to cut off the end portion after pulling the cable to ensure an undamaged end. Use sharp, high quality tools. When the various layers are removed, cuts should extend only partially through the layer. Be careful not to cut completely through and damage conductor strands when removing cable insulation. A useful technique for removing polyethylene cable insulation is to use a string as the cutting tool.

When penciling is required (unnecessary for molded rubber devices), a full smooth taper is necessary to eliminate the possibility of air voids. Semiconductive layer(s) and the resulting residue must be completely removed. Two common methods are to use abrasives and solvents.

A 120-grit abrasive is best for removing semiconductor and residue. A 120-grit abrasive is fine enough for the high-voltage interface, yet coarse enough to remove semiconductor residue without "loading up" the abrasive cloth. Use nonconductive abrasive grit. *Do not use* emery cloth or any other abrasive that contains conductive particles, because these particles could embed themselves in the cable insulation.

Be careful not to abrade the insulation below the minimum specified for the device when using an insulation diameter-dependent device such as molded rubber.

A nonflammable cable cleaning solvent is the preferred method. Avoid using solvents that leave residues. *Do not use* excessive amounts of solvent or saturate the semiconductive layers, because doing so will render the layers nonconducting. Know the solvent being used, and avoid toxic solvents that are hazardous to health.

There are cable preparation kits available that contain 120-grit nonconductive abrasive cloth saturated with nonflammable, nontoxic 1–1–1 trichloroethane, which make these kits self-contained for field use (see Figure 3–28).

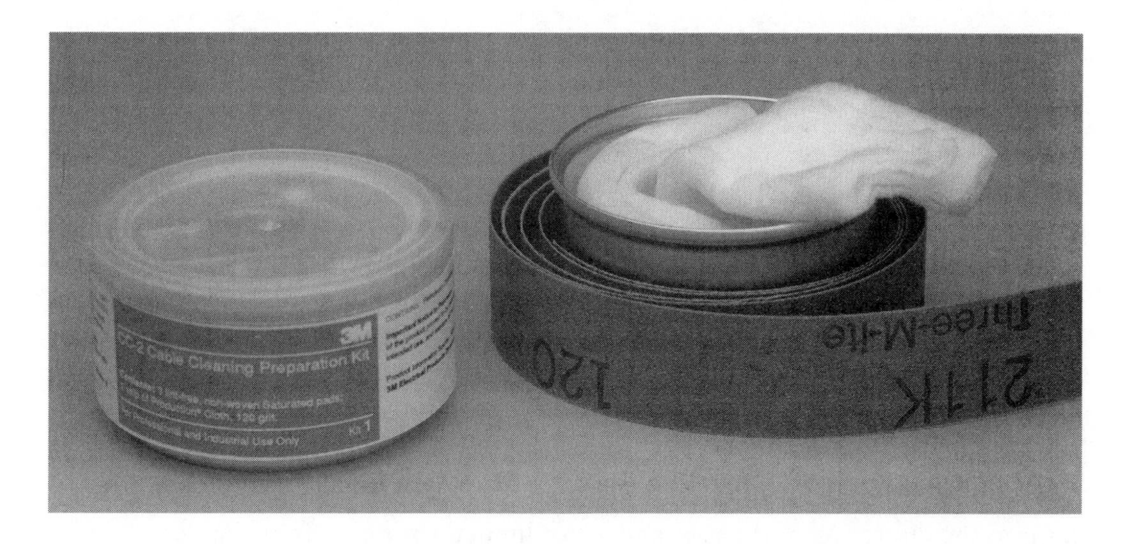

Figure 3–28 Cable preparation kit.
(Courtesy of 3M Electrical Products Division)

After cable surfaces have been cleaned, the recommended practice is to reverse wrap a layer of vinyl tape, adhesive side out, to maintain the cleanliness of the cable.

High-quality products usually include detailed installation instructions. These instructions should be followed. A suggested technique is to check off steps as they are completed. Good instructions alone do not qualify a person as a cable splicer. Some manufacturers offer hands-on training programs designed to teach proper installation of their products. It is highly recommended that inexperienced splice and termination installers take advantage of such programs.

Joining Conductors. After the cables are completely prepared, the rebuilding process begins. The first step is to reconstruct the conductor with a suitable connector. A suitable connector for high-voltage cable splices is a compression-type device (see Figure 3–29). *Do not use* mechanical-type connectors, such as split-bolts. Connector selection is also based on the conductor material.

Generally, copper conductors should be joined with connectors that are marked "CU" or "AL-CU." Aluminum should be joined only with connectors marked "AL-CU." "AL-CU" may come preloaded with no-oxide paste, which is used to break down the aluminum oxide on the surface. If the connector is not preloaded, no-oxide paste must be applied.

It is recommended that a UL-listed connector be used that can be applied with a common crimping tool (see Figure 3–30). This connector should be tested and

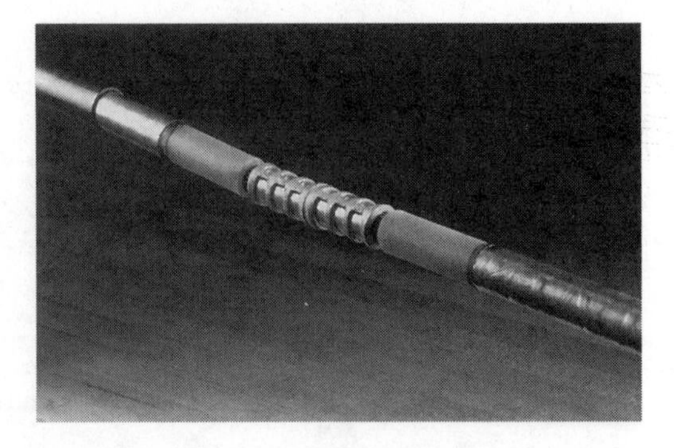

Figure 3–29 Compression connectors. *(Courtesy of 3M Electrical Products Division)*

approved for use at high voltage. In this way, the choice of the high-voltage connector is at the discretion of the user and is not limited by the tools available.

Reinsulating. Perhaps the most commonly recognized method for reinsulating is the tape method. Tape is not dependent upon cable types and dimensions, has a history of dependable service, and is generally available. Wrapping tape on a high-voltage cable can be time-consuming and error-prone because the careful buildup of tape requires accurate half-lapping and constant tension to reduce built-in air voids (Figure 3–31).

New technology has made available linerless splicing tapes. These tapes reduce both application time and error. Studies have shown time savings of 30 to 50 percent are possible because there is no need to stop during taping to tear off the liner. This also allows the splicer to maintain a constant tape tension, thereby reducing the possibility of taped-in voids. Tape splice kits that contain all the necessary tapes,

Figure 3–30 Compression tool. *(Courtesy of 3M Electrical Products Division)*

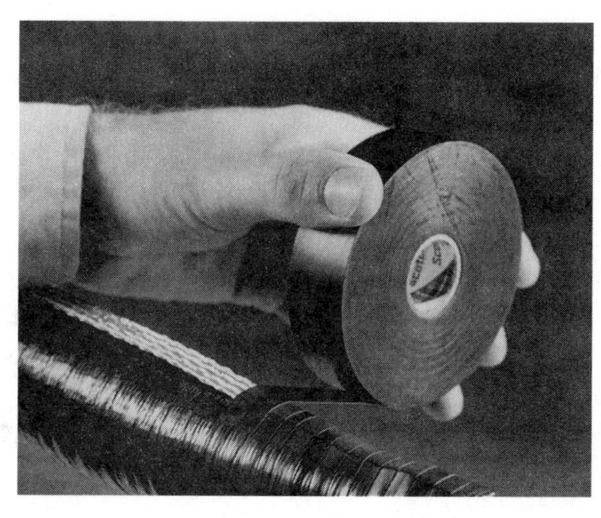

Figure 3–31 Half-lapping onto a splice. *(Courtesy of 3M Electrical Products Division)*

along with proper instructions, are available, and they make ideal emergency splice kits.

Another method for reinsulating uses molded rubber technology. These factory-made splices are designed to be human engineered for the convenience of the installer. In many cases, these splices are also factory tested and designed to be installed without special installation tools.

All molded rubber splices use EPR as the reinsulation material (see Figure 3–32). EPR must be cured during the molding process using either peroxide or

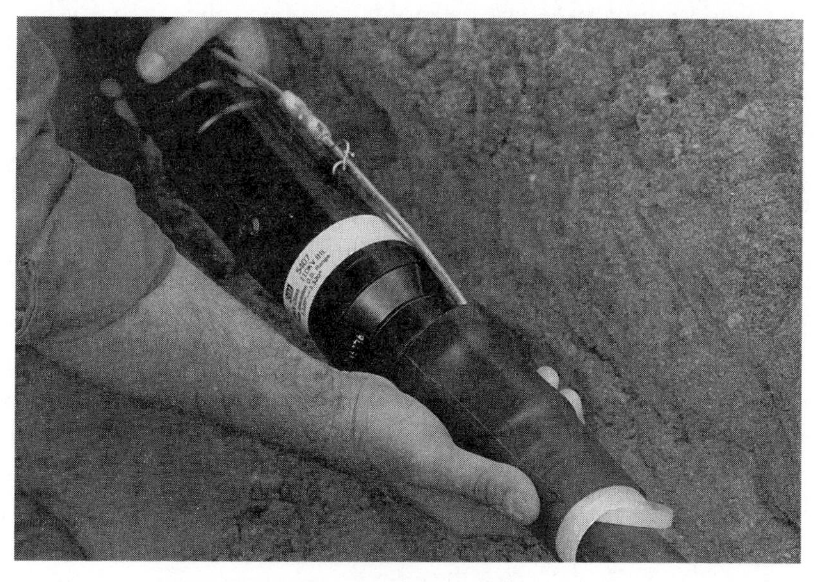

Figure 3–32 Molded rubber splice kit.
(Courtesy of 3M Electrical Products Division)

sulphur. Peroxide cures develop a rubber with maximum flexibility for ease of installation and, more important, provide an excellent long-term live memory for lasting reliable splices.

Reshielding. The cable's strand shield and insulation shield system must be rebuilt when constructing a splice. The tape and molded rubber methods are used in a similar fashion as was outlined for the reinsulation process.

For a tape splice, the cable strand shielding is replaced by a semiconductive tape. This tape is wrapped over the connector area to smooth the crimp indents and connector edges.

The insulation shielding system is replaced by a combination of tapes. Semiconductor material is replaced with the same semiconducting tape used to replace the strand shield, as shown in Figures 3–33 and 3–34.

The cable's metallic shield is generally replaced with a flexible woven mesh of tin-plated copper braid. This braid is for electrostatic shielding only and is not designed to carry shield currents. For conducting shield currents, a jumper braid is installed to connect the cable's metallic shields. This jumper must have an ampacity rating equal to that of the cable's shields (see Figures 3–35 and 3–36).

For a rubber molded splice, conductive rubber is used to replace the cable's strand shielding and the semiconductive portion of the insulation shield system (see Figure 3–37). Again, the metallic shield portion must be jumpered with a metallic component of equal ampacity.

A desirable design parameter of a molded rubber splice is that it be installable without special installation tools. To accomplish this, very short electrical interfaces are required. These interfaces are attained through proper design shapes of the conductive rubber electrodes.

Laboratory field plotting techniques show that the optimum design can be obtained using a combination of logarithmic and radial shapes.

Rejacketing. Rejacketing is executed in a tape splice using a rubber splicing tape overwrapped with a vinyl tape (see Figure 3–38).

In a molded rubber splice, rejacketing is accomplished by proper design of the outer semiconductive rubber, effectively resulting in a semiconductive jacket (see Figure 3–39). When a molded rubber splice is used on internally shielded cable (such as ribbon shield or drain wire shield), a shield adapter is used to seal the opening that results between the splice and the cable jacket.

Step-by-Step Procedure. The following procedure is provided by the Rome Cable Corporation and refers to Figure 3–40.

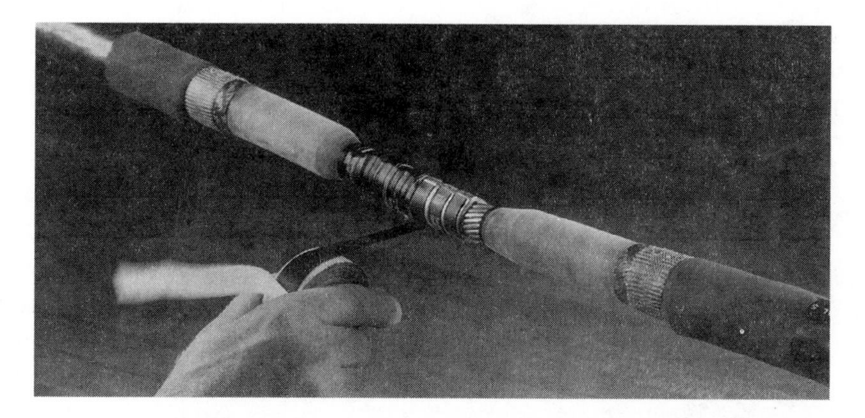

Figure 3–33 Replacing the cable strand shield with semiconducting tape.
(Courtesy of 3M Electrical Products Division)

Figure 3–34 The insulation semiconducting shield.
(Courtesy of 3M Electrical Products Division)

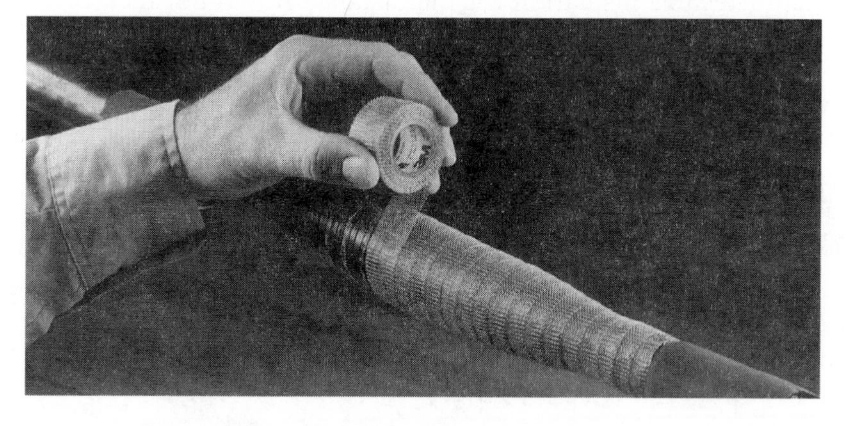

Figure 3–35 Replacing the electrostatic shield.
(Courtesy of 3M Electrical Products Division)

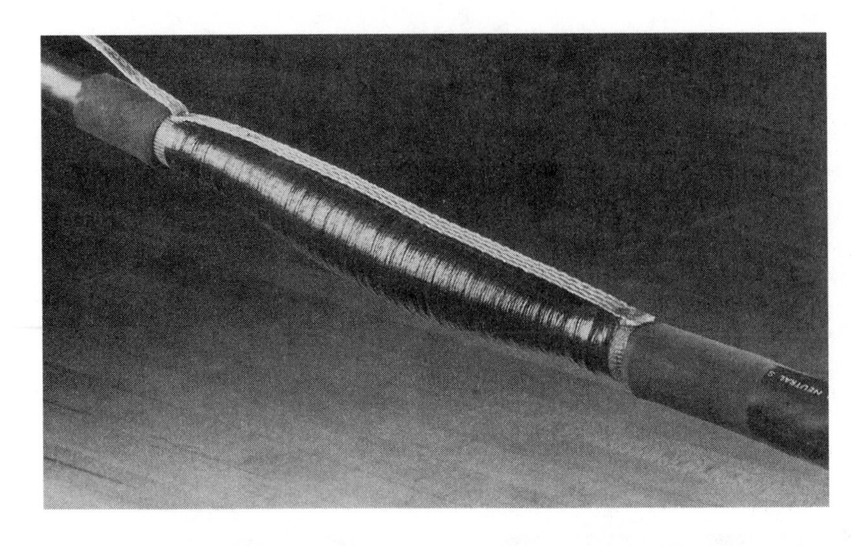

Figure 3–36 Jumper braid installed to carry shield current. *(Courtesy of 3M Electrical Products Division)*

1. Train cable ends into position. Cut cables squarely at centerline of splice.
2. Measure distance *A* on each end and mark cables.
3. Remove cable jacket and underlying separator tape from each cable to point marked.
4. Remove metal shielding.
   a. For tape-shielded cable, remove to where ¾ inch of this material protrudes beyond the jacket.
   b. For wire-shielded cable, unwind wires, bend over jacket, and twist them into a common conductor.
5. Remove the semiconducting layer to the point where ¾ inch of this material protrudes beyond the metal shielding.
6. Temporarily tie the shielding in place with tape.
7. Remove cable insulation for the distance *B*. Do not nick or cut the conductor.
8. Remove strand shielding, which should be cut cleanly at the junction with the conductor.

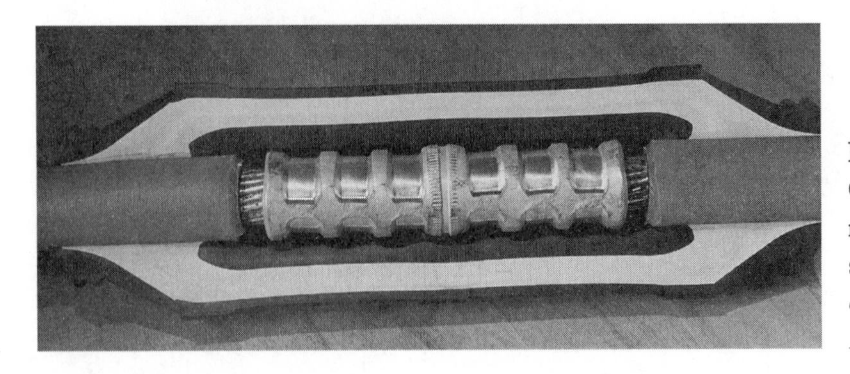

Figure 3–37 Cutaway view of a molded rubber splice. *(Courtesy of 3M Electrical Products Division)*

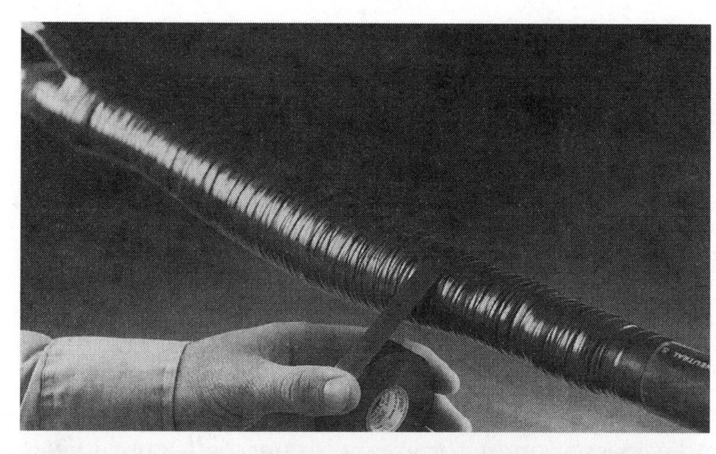

Figure 3–38 Rejacketing using electrical tape. *(Courtesy of 3M Electrical Products Division)*

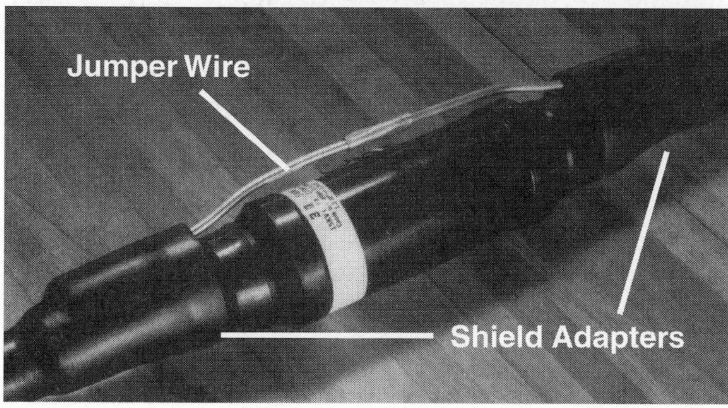

Figure 3–39 Rejacketing using a molded rubber splice. *(Courtesy of 3M Electrical Products Division)*

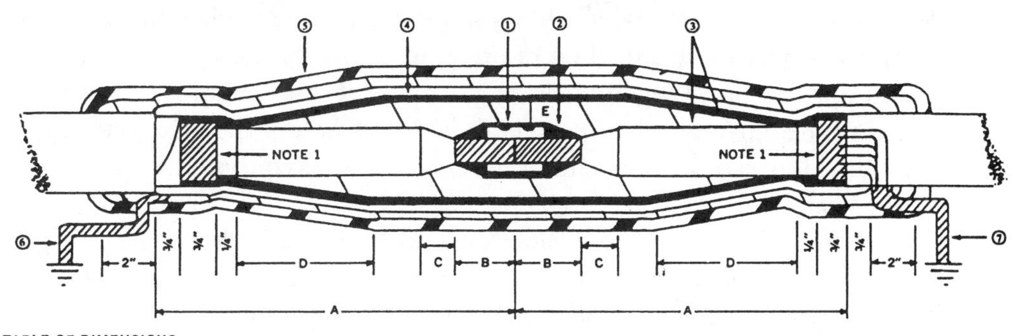

TABLE OF DIMENSIONS

Dimen.	Rated kV		
	5	15	25
A	One-Half Connnector Length Plus 6"	9¾"	13"
B	One-Half Connnector Length Plus ½"	½"	½"
C	¾"	1¼"	2"
D	3½"	6"	7½"
E	3/16"	3/8"	9/16"

① Compression Connector
② Conducting Tape
③ Insulating Tape
④ Tinned Copper Mesh
⑤ Vinyl Tape for Plastic Jacketed Cable
Neoprene Tape, for Neoprene Jacketed Cable
⑥ Ground Lead
⑦ Twisted Drain Wire Ground Lead

NOTE

1. Remove all traces of semiconducting material to this point.

Figure 3–40 Splicing 5, 15, and 25 kV 1/C shielded PE, XLP, and EPR insulated cables. *(Courtesy of Rome Cable Corporation)*

9. Pencil insulation of each cable end for the distance *C*. Smooth with file or abrasive cloth.
10. Inspect the area at the end of the pencil for protruding fibers of strand shield.
11. Pencil and smooth exposed edges of cable jacket.
12. Install connector per manufacturer's instructions. Smooth any major roughness with a file or an abrasive cloth.
13. Wipe jacket, insulation, and connector with a clean cloth, preferably one that has been moistened with a suitable solvent, such as trichloroethane or trichlorethylene.
14. Fill connector indents with pieces of conducting shielding tape.
15. Cover connector and exposed conductor area with one half-lapped layer of conducting shielding tape.
16. Apply insulating tape to a wall thickness of *E* over the connector shield. Employ level-wind technique and apply in successive half-lapped layers. The ends of the applied insulation should be tapered for a distance *D* as shown and should end approximately ¼ inch from the edge of the cable insulation shield.
17. Holding the semiconducting tapes, remove the tape applied in Step 6.
18. Apply a half-lapped layer of conducting shielding tape over the insulating tape. Starting at the very edge of the metal shielding, cover the cable semiconducting layer, the insulating tape, and the semiconducting layer on the opposite side of the splice.
19. Holding the cable metal shielding, remove the tape applied in Step 6.
20. Apply copper mesh shielding braid. Start at one of the cable metal shields and apply a half-lapped layer of the braid across the splice to the other metal shield overlapping this metal shielding. Solder the tape to the metallic shielding tape at each end of the braid wrap, avoiding the use of excessive heat.
21. With tape-shielded cable, solder a ground lead to the braid, avoiding the use of excessive heat. Solder block this lead or use a solid conductor.
22. Apply jacket tapes. Scrape the jacket clean for 3 inches on each side of splice.
    a. For Neoprene-jacketed cables, apply three half-lapped layers of insulating tape across the splice. Extend the insulating tape wrap at least 2 inches beyond the braid at each end of the splice area.
    b. For plastic-jacketed cables, apply three half-lapped layers of insulating tape across the splice. Extend the insulating tape wrap at least 2 inches beyond the braid at each end of the splice area. Next, apply two half-

lapped layers of vinyl electrical tape over the insulating tape, extending 1 inch onto the cable jacket.

23. Insert shield ground between layers of jacketing tape.

Other Splice-Insulation Methods

A variety of splice insulation methods are available for plastic insulation in addition to the tape wrapping and the molded kit types, which have been discussed. The following paragraphs briefly describe these additional methods.

Poured Epoxy/Resin Filled. This type of insulation (see Figure 3–41) is particularly useful when a splice on a nonshielded cable must be moisture proof. The following steps are executed after the conductors are properly spliced using one of the previously described methods.

1. Snap the two-piece mold body together over the splice area.
2. Seal the ends of the mold body using electrical tape.
3. Insert the funnel(s) into the mold body.
4. Prepare the resin and pour it slowly into the funnel until filled.

Figure 3–41 Resin splice insulating kit.
(Courtesy of 3M Electrical Products Division)

5. After the resin sets, remove the funnel and the mold body.
6. Dress the resin splice as required.

Heat Shrink. Heat-shrink tubing is another excellent choice for replacement of the insulation and jacket on a low-voltage cable splice (see Figure 3–42). The application of heat shrink is quick and simple and provides a dependable covering. The procedure, which follows, is straightforward.

1. Slip the heat-shrink tubing over one of the ends to be spliced.
2. Splice the conductor as described under low-voltage and nonshielded splicing.
3. Position the heat shrink over the splice and apply heat from a portable torch to shrink the tubing until it conforms to the splice.

The principal disadvantage to heat-shrink material is that it requires heat. Use extreme care when working with open flame.

Cold Shrink. A recent addition to insulating splices is cold-shrink materials. Cold-shrink tubing is a prestretched, tubular rubber sleeve with a removable, collapsible core. When the tubing is applied in the field, the core material is removed, thereby allowing the tubing to collapse, or shrink, onto the splice. The procedure is as follows:

1. Slip the cold-shrink tubing over one of the ends to be spliced.
2. Splice the conductor as described under low-voltage and nonshielded splicing.

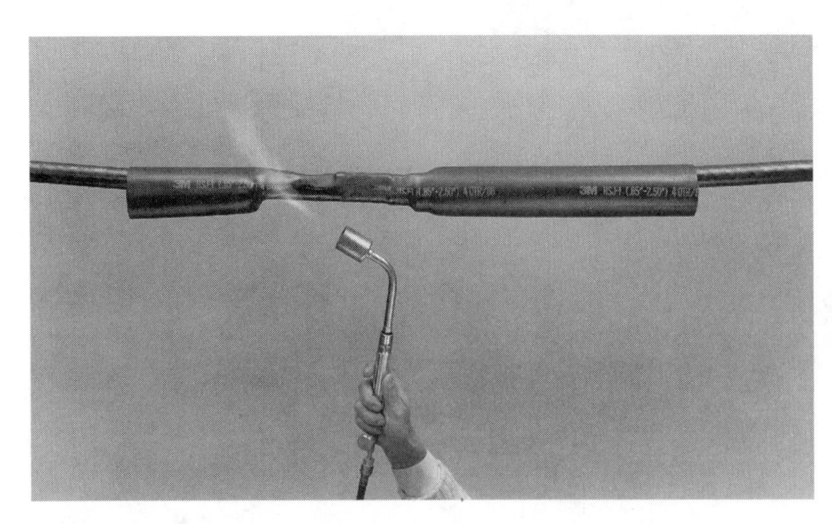

Figure 3–42 Heat shrink being applied to a splice. *(Courtesy of 3M Electrical Products Division)*

3. Apply a layer of electrical tape.
4. Position the cold-shrink tubing over the splice and start removing the collapsible core; the rubber tubing will collapse onto the splice.

Cold-shrink tubing creates a heat-, chemical-, and moisture-resistance covering. Figures 3–43, 3–44, and 3–45 show the procedure for applying cold-shrink tubing.

Lead-Sheathed Cable. Medium-voltage lead cable splicing kits are available, which make the task of splicing lead much simpler. Their principal disadvantage is their cost. The general procedure for splicing lead using such a kit is as follows:

1. Slide the preformed lead tube over one end of the two cables.
2. Complete the conductor splice as previously described. The conductor may be soldered when splicing lead cable, but this is not required.
3. Layer tape onto the splice.
4. Slide the preformed tube over the splice and seal both ends with solder.
5. Fill the tube with the resin compound supplied with the kit.
6. Plug or solder the fill holes.
7. Allow the splice to cool completely before moving it.

Transition kits for splicing plastic cable to lead-sheathed cable are also available.

TERMINATIONS

Overview

The termination of a cable has much in common with splicing. A cable termination generally needs to provide the following:

- Low electrical resistance connection from the cable to the equipment to which the cable is being terminated.
- Adequate insulation from all conductors.
- Protection from the entrance of moisture, chemicals, and other harmful environmental effects.

The level of sophistication required for a given termination depends on the voltage level, the current requirements, the size of the cable, and the type of insulation. The following paragraphs describe several different types of terminations and provide some information to execute them.

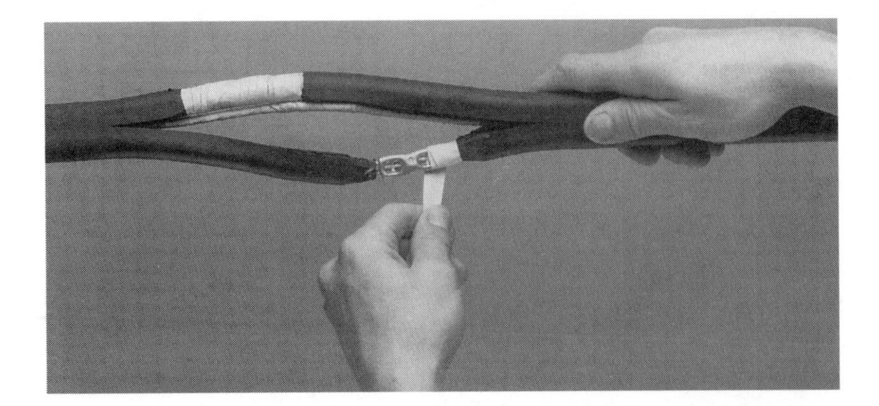

Figure 3–43 Completing the conductor splice prior to installing the cold shrink. *(Courtesy of 3M Electrical Products Division)*

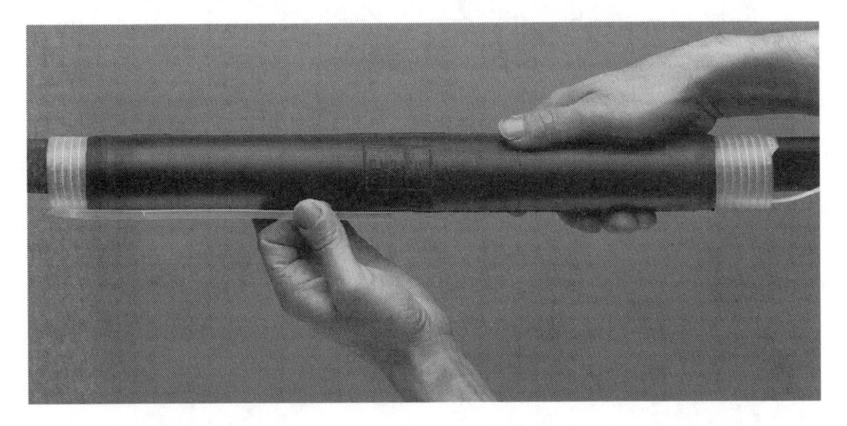

Figure 3–44 Positioning the cold shrink in starting to remove the collapsible core. *(Courtesy of 3M Electrical Products Division)*

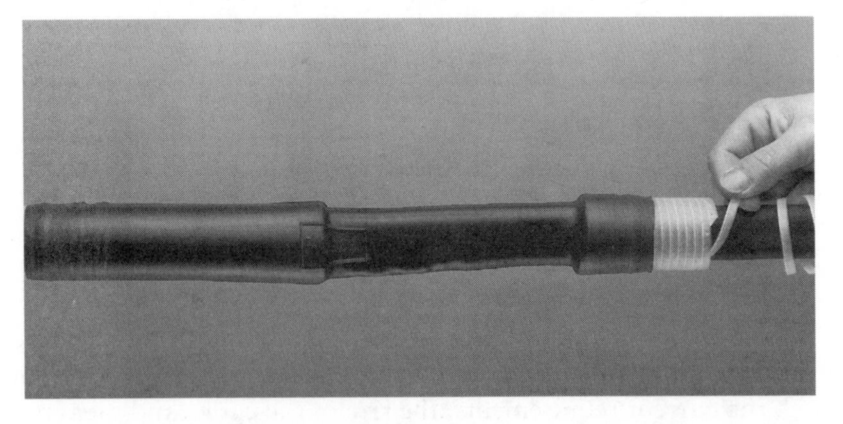

Figure 3–45 Collapsible core almost removed from one end. *(Courtesy of 3M Electrical Products Division)*

Low Voltage

Terminal Binding. Figure 3–46 shows a typical terminal binding type of termination. This type of connection is typically used in branch circuit-to-fixture types of connections and control cable terminations. The wire sizes vary but are usually #10 AWG and smaller. Solid or stranded wire may be terminated in this way, but solid wire is preferable. If stranded wire is used, one of the following methods may be used.

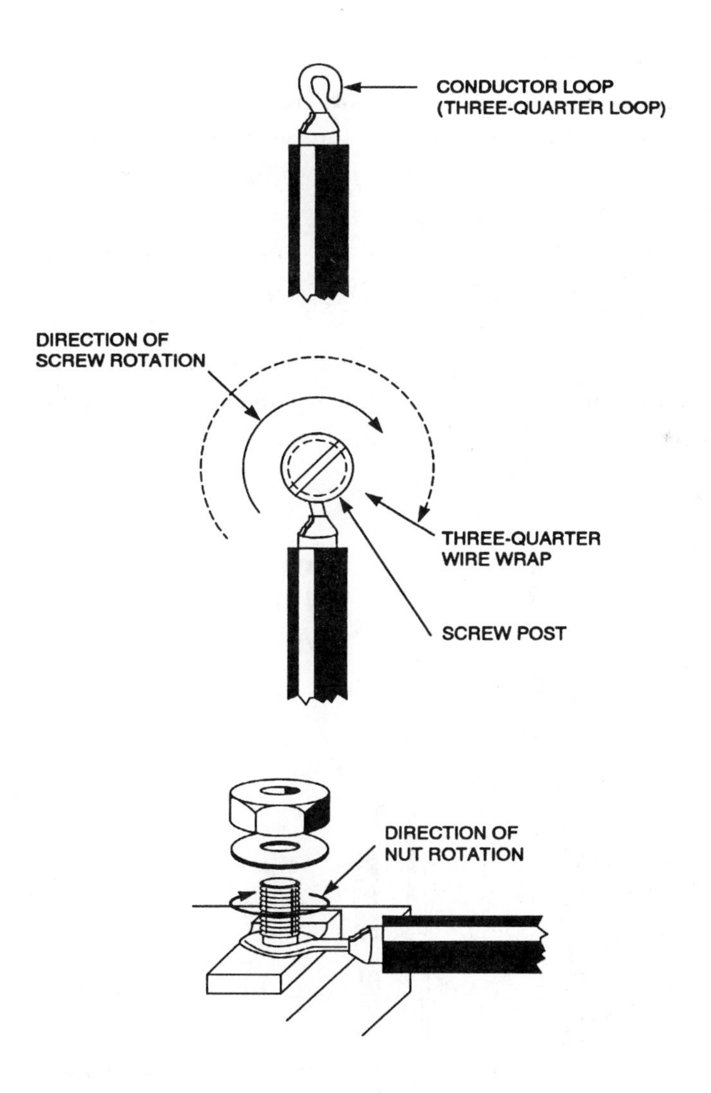

Figure 3–46 Typical terminal binding methods.

Low-Voltage Terminal Lugs. When stranded wire is used, low-voltage crimp-on terminal lugs are the preferred way for terminating low-voltage wires in sizes #10 AWG and smaller. Figures 3–47 and 3–48 show typical lugs and their applications to the wire. Lugs of this type are squeezed on using a hand crimper. Figure 3–49 shows a typical hand crimper.

Note that these types of terminal lugs may be used to allow the use of stranded wire where solid wire is normally employed. For example, solid wire (NM or NMC) is normally used in electrical junction boxes. If stranded wire is used because of its flexibility or higher conductivity, crimp-on terminal lugs must be applied to the wire first. The lug can then be inserted under the screw head or nut and secured.

Compression Lugs. Compression lugs are used for higher current, voltage, and wire-size applications. The compression lug benefits from application pressure all the way around the wire, as opposed to the smaller crimped type of connection. Figures 3–50 and 3–51 show compression-type lugs and tools, respectively. Figure 3–52 shows examples of various types of compression lugs after they have been applied to wire. This type of lug is used typically on #8 AWG and larger wire.

Lug compressors are designed to fit a certain type of lug, but some compressors are equipped with interchangeable dies to accommodate various types of lugs. Manufacturer's instructions and reference tables *must* be followed to ensure that the correct die is used for the lug being applied.

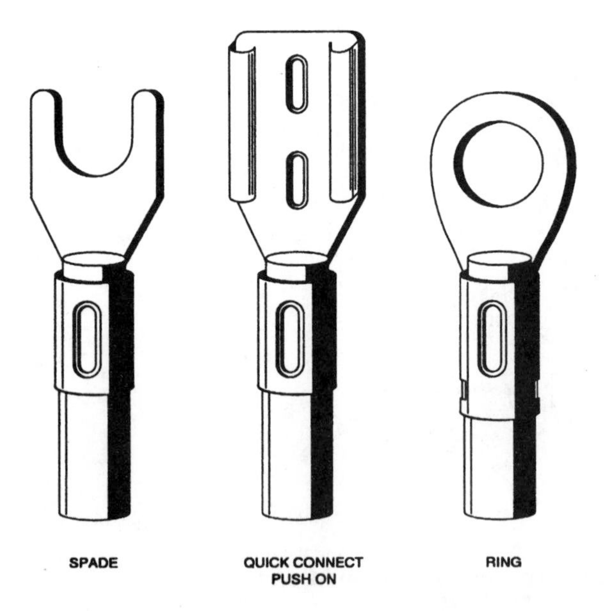

Figure 3–47 Typical low-voltage crimp-on terminal lugs.

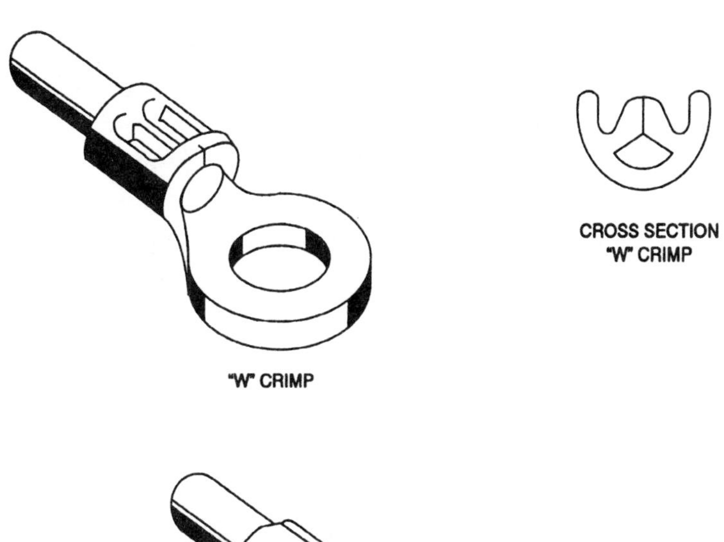

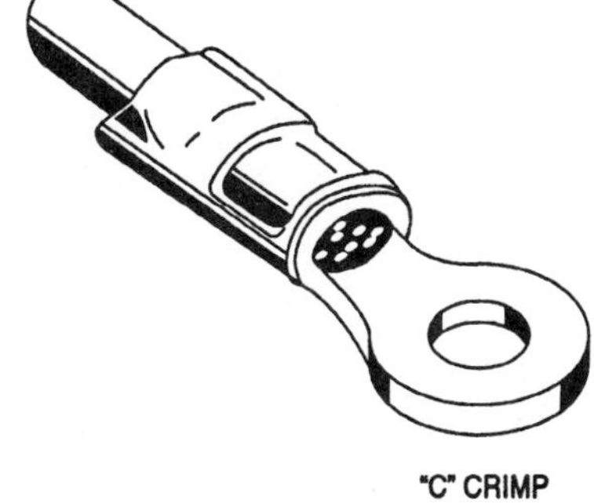

Figure 3–48 Low-voltage terminal lugs after crimping.

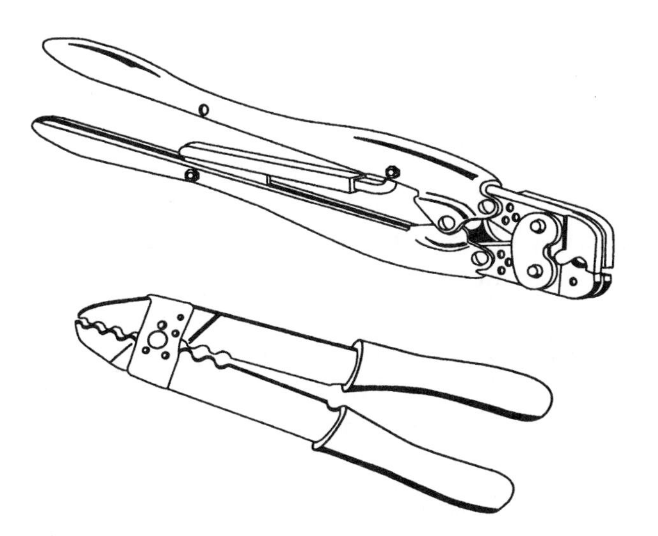

Figure 3–49 Hand-operated crimping tools.

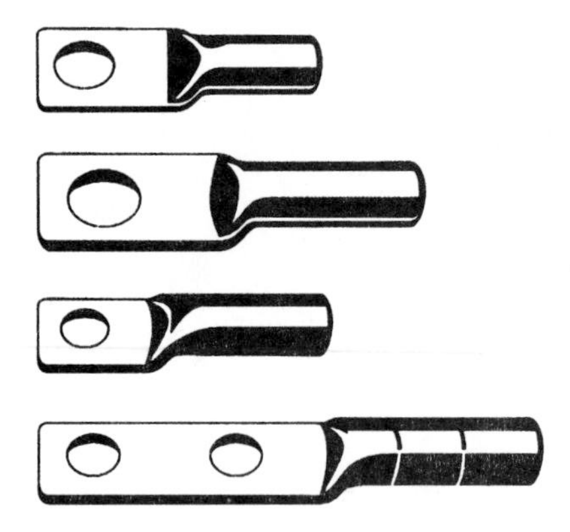

Figure 3–50 Compression lugs.

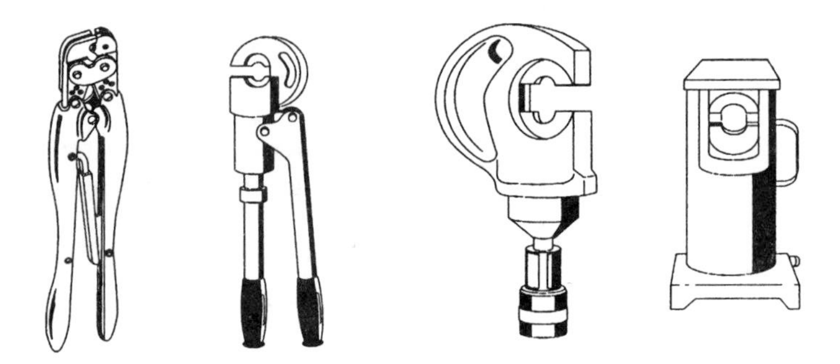

Figure 3–51 Mechanical and hydraulic compression tools.

Bolted Terminal Lugs. Bolted terminal lugs are used for virtually all voltage and current ratings. The principal advantage of this type of termination method is the ease of application and removal. Consequently, bolted terminal lugs are commonly applied in applications in which frequent wiring changes may be required.

Bolted terminal lugs are subject to loosening due to heat, vibration, or chemical action. As a result, the lug must be applied carefully with the proper amount of torque, administered by a torque wrench. Figure 3–53 shows various types of bolted terminal lugs, and Figure 3–54 shows the application of a bolted terminal lug using a torque wrench.

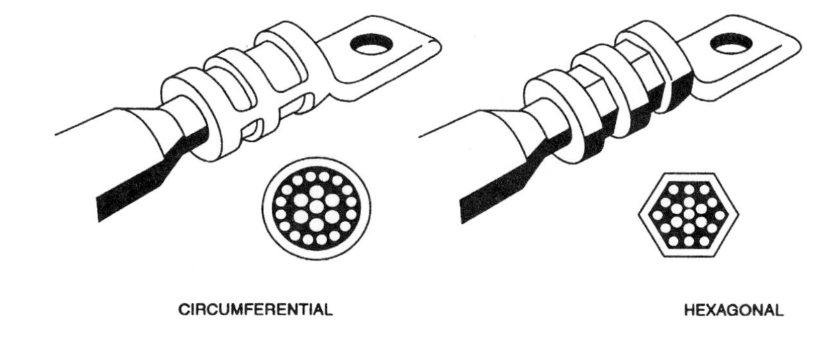

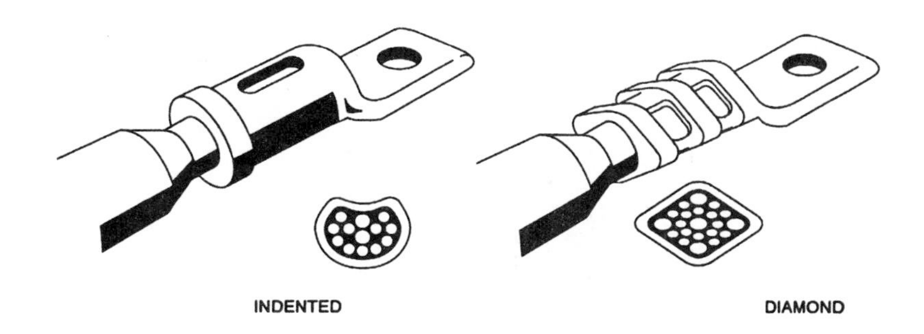

Figure 3–52 Compression configuration for high-voltage terminal lugs.

Procedure. One of the most important steps in any cable termination is the restoration of the insulation system. The following procedure, furnished by Rome Cable Corporation, gives a typical example for a complete termination (see Figure 3–55).

1. Train cable end into position. Mark and cut squarely.
2. Measure a distance ¾ inch + depth of bore in terminal lug from cable end. Remove the cable jacket, insulation, and strand-shielded tape (if present) to this point.
3. Pencil the cable jacket and insulation to a distance of 1 inch.
4. Smooth the surface of pencilled area with a file or cloth. Be sure the strand shielding is cut *cleanly* at the end of the pencil and that no fibers remain.
5. Install the terminal lug.
6. Wipe the uppermost 2 inches of cable jacket, pencilled area, conductor, and terminal lug with a clean cloth, preferably moistened with a solvent such as trichloroethane or trichlorethylene.

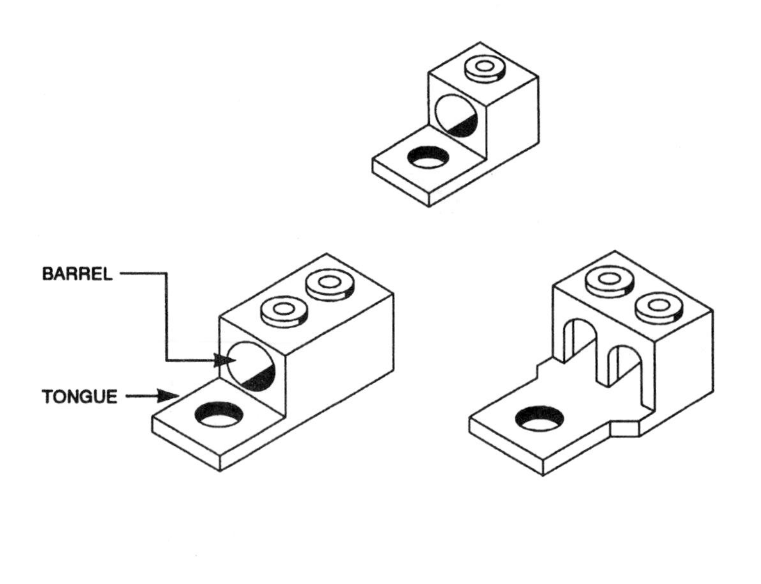

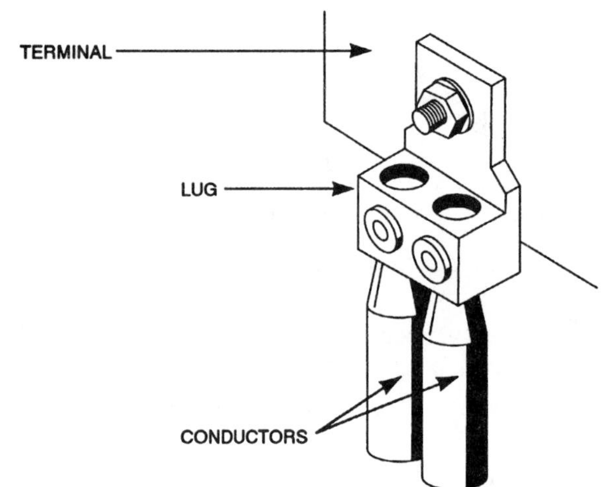

Figure 3–53 Bolted terminal lugs.

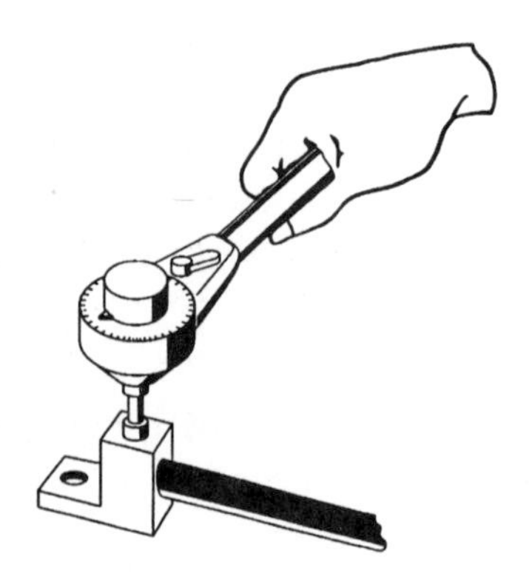

Figure 3–54 Torque wrench for application of a bolted terminal lug.

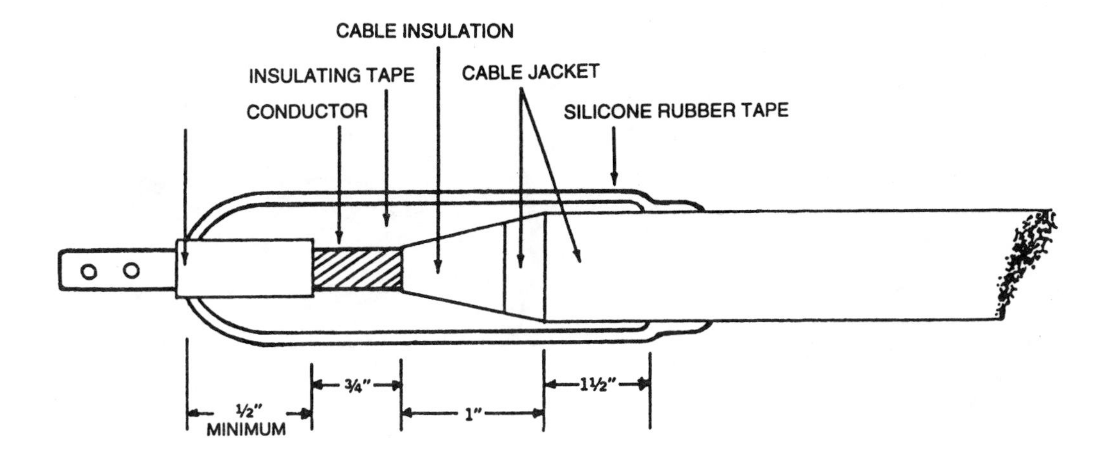

Figure 3–55 Termination for unshielded 0–5 kV cable PE, XLP, and EPR insulated cables.
(Courtesy of Rome Cable Corporation)

7. Fill the indents in the terminal lug with filler tape.
8. Apply insulating tape. Start at the area of exposed conductor and build to the level of the lug shoulder. Apply successive half-lapped layers of tape. Build a tape wall from the broad end of the pencil to a point not less than ½ inch up the terminal lug. When this wall reaches a diameter equal to the diameter of the cable jacket, complete the taping by applying an additional three half-lapped layers of tape extending at least 1½ inches onto the jacket.
9. For maximum protection from water and atmospheric contamination outdoors, overwrap the insulating tape with two half-lapped layers of self-fusing silicone rubber tape.

Medium Voltage

The termination of medium- and high-voltage cable differs from low-voltage cable principally in the method used to alleviate high-voltage electrical stress. Under normal circumstances, the electric field in a shielded cable is uniform along the axis of the cable and varies along the radius of the cable. The greatest voltage gradient is located toward the center (conductor) of the cable. This condition is shown in Figures 3–56 and 3–57.

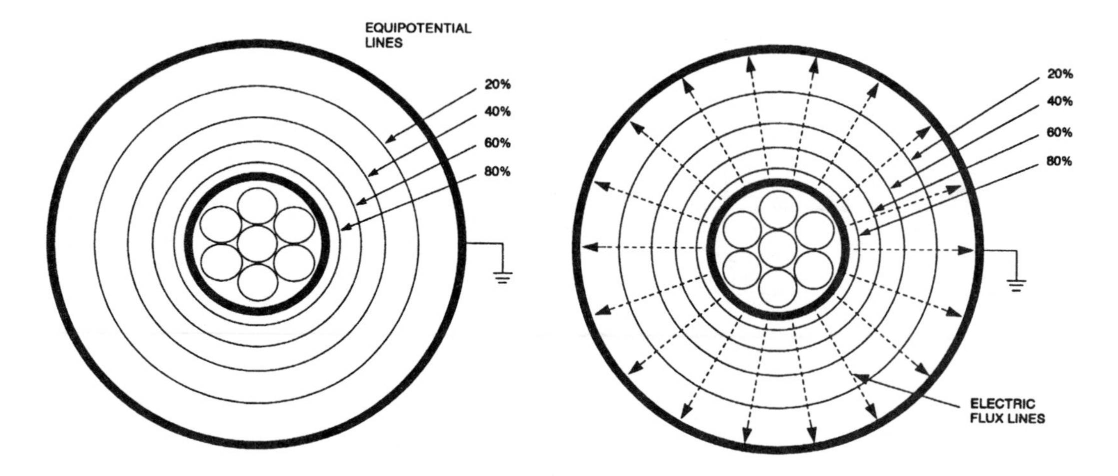

Figure 3–56 Variation of electric flux in a shielded cable. *(Courtesy of 3M Electrical Products Division)*

Figure 3–57 Flux lines in a shielded cable. *(Courtesy of 3M Electrical Products Division)*

When a cable is terminated, the semiconductor shield is removed for some distance from the end of the cable. This causes the electric flux lines to concentrate at the end of the semiconducting shield, as shown in Figure 3–58.

There are two ways to reduce the electrical stress: geometric stress control and capacitive stress control. In the geometric method, shown in Figure 3–59, the diameter of the termination is expanded and reduces the stress by enlarging the radius of the shield. This method produces the familiar stress cone at the end of many medium-voltage terminations.

Capacitive stress control methods use materials with different dielectric constants at the termination (see Figure 3–60). In this technique, materials with progressively lower dielectric constants *(K)* are wrapped around the cable. This change in dielectric constant allows the flux lines to distribute evenly, thereby reducing the stress to tolerable levels.

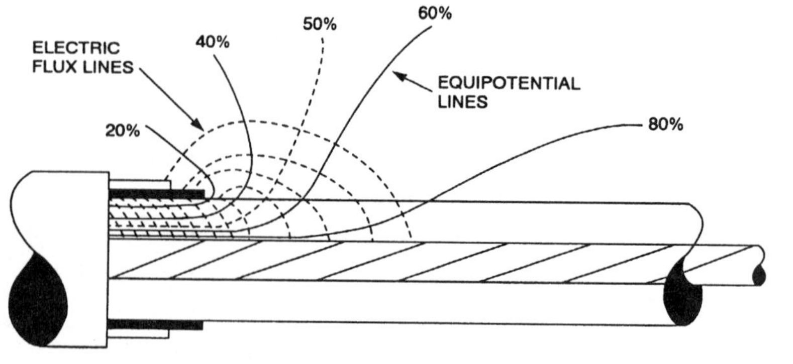

Figure 3–58 Flux lines in shielded cable cut for termination. *(Courtesy of 3M Electrical Products Division)*

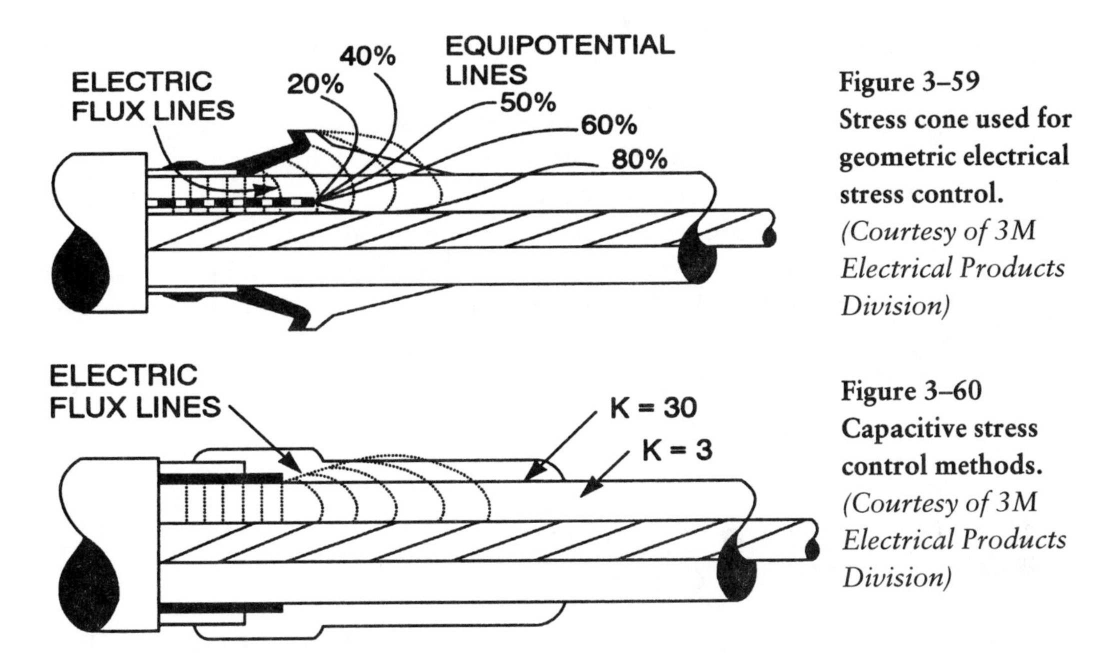

Figure 3–59
Stress cone used for geometric electrical stress control.
(Courtesy of 3M Electrical Products Division)

Figure 3–60
Capacitive stress control methods.
(Courtesy of 3M Electrical Products Division)

High-voltage termination manufacturers provide medium-voltage termination kits that contain all the necessary materials to terminate, insulate, and seal the termination. Figure 3–61 is an example of a medium-voltage termination.

Terminal Blocks

Terminal blocks, as shown in Figure 3–62, are assemblies that are used for splicing and terminating low-voltage wire used primarily in control applications.

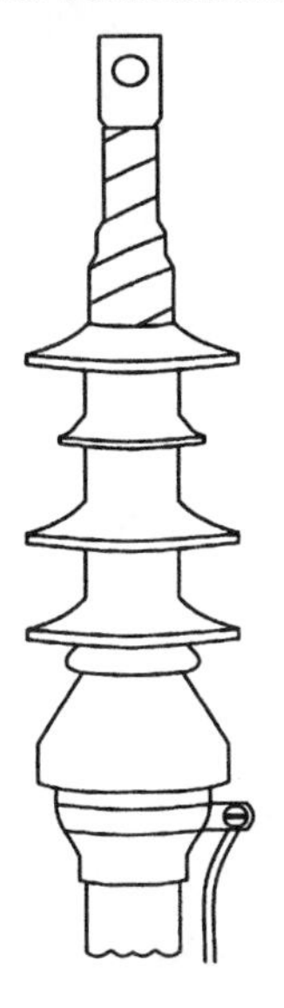

Figure 3–61 Medium voltage terminations.
(Courtesy of 3M Electrical Products Division)

Figure 3–63 shows the general configuration for connecting an external cable to an internal wiring system.

Terminal blocks may be used to connect either solid or stranded wire. When solid wire is used, the connection may be an eye type of connection, as shown in Figure 3–64, the tubular pressure screw-type connector, as shown in Figure 3–64a, or the screw clamp-type connector, as shown in Figure 3–64b.

When stranded wire is used, the wire is first terminated in a crimp- or compression-type lug as described earlier. The lug is then placed under the screw-type connector (see Figure 3–64c) and the screw is tightened. To ensure a secure and lasting connection, split-ring lockwashers or self-locking screw heads should be used on the terminal block.

TERMINATING AND SPLICING COMMUNICATIONS CABLE

As mentioned earlier in this chapter, cable splicing should be avoided whenever possible. Terminations, in contrast, cannot be avoided. Misapplied connectors can be sources of problems in premises cabling. Particularly with Category 5 UTP

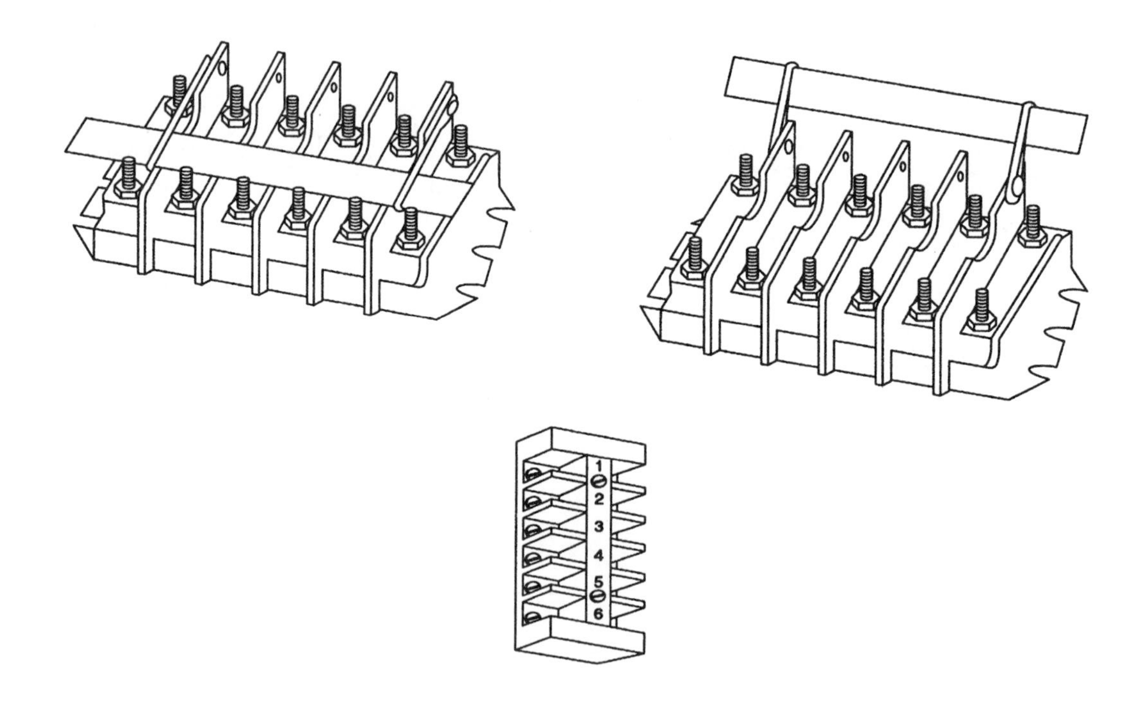

Figure 3–62 Various types of terminal blocks with marker strips.

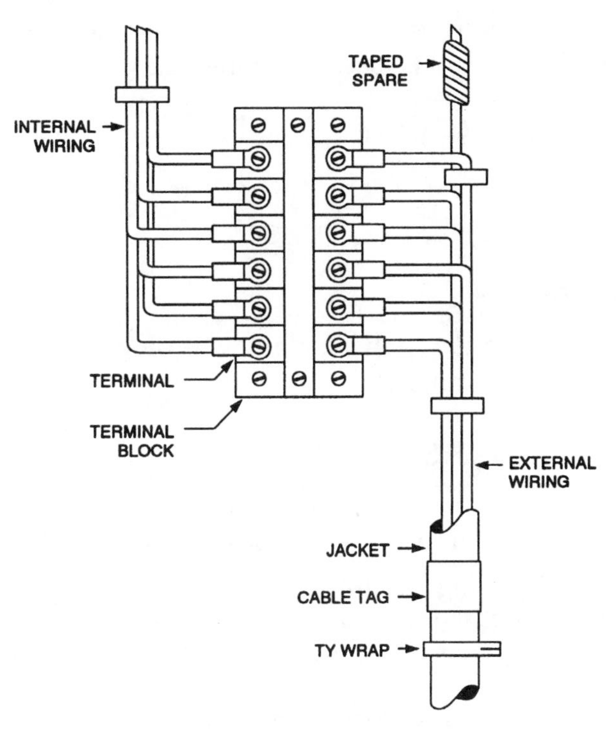

Figure 3–63 Wiring connections to terminal block.

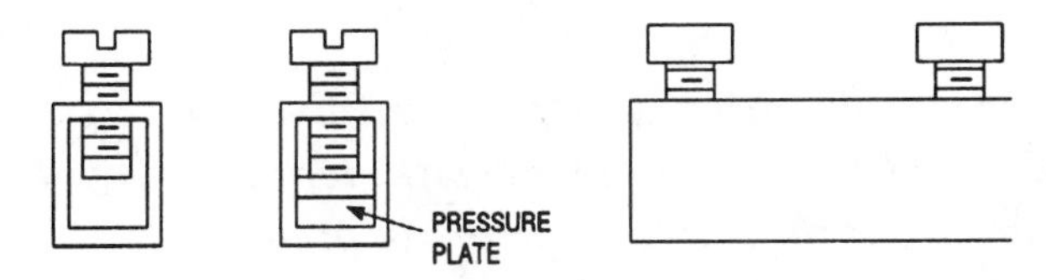

a. TUBULAR PRESSURE SCREW-TYPE CONNECTOR

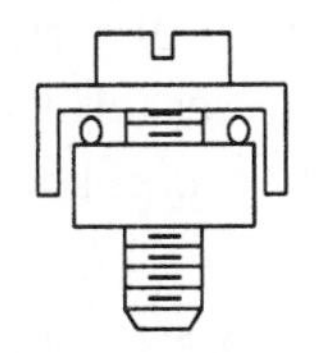

b. SCREW CLAMP-TYPE CONNECTOR

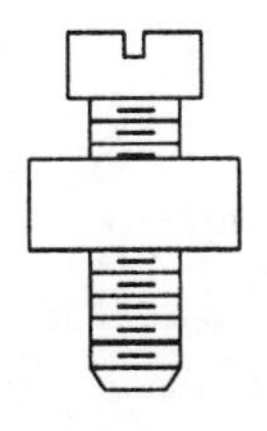

c. SCREW-TYPE CONNECTOR

Figure 3–64 Typical screw-type connectors for terminal blocks.

and fiber, careless application can degrade performance considerably. Because most network problems occur in cables and connectors, special care must be used in terminations.

The following information was taken verbatim from *Premises Cabling* by Donald J. Sterling, Jr.

Terminating Modular Connectors

Figure 3–65 shows the basic procedure for applying a modular plug to a connector. The general termination procedure is simple:

- Strip away the cable jacket to expose the individual pairs.
- Untwist the pairs just enough to insert them into the plug—no more than ½ inch for Category 5 cable and 1 inch for Category 4.
- Insert the cables into the plug, being careful to use the correct order for the wiring pattern being used.
- Cycle the crimping tool.
- Inspect the termination. The clear plastic of the plug allows you to inspect the termination area and check that the pairs are in the right order.

> *The importance of carefully inspecting each termination cannot be overemphasized. Careless or inadvertent terminations are a major source of problems and errors in premises cabling.*

Figure 3–66 shows a typical termination procedure for terminating a modular jack with 110-style punch-down blocks. The wires are carefully positioned in the contacts (taking care to limit untwisting of pairs) and then pushed into place with a punch-down tool.

Terminating a cable in a punch-down block is fairly straightforward. The wire is positioned in the IDC contact and the punch-down tool is applied to seat it in the contact. The tool also cuts off the end of the wire.

Here are some things to pay attention to:

- Remove as little of the outer jacket as possible.
- Maintain the twist on the cables as close to the connector as possible. *With Category 5 UTP, do not untwist more than 0.5 inch.* For Category 4 cable, the limit is 1 inch.

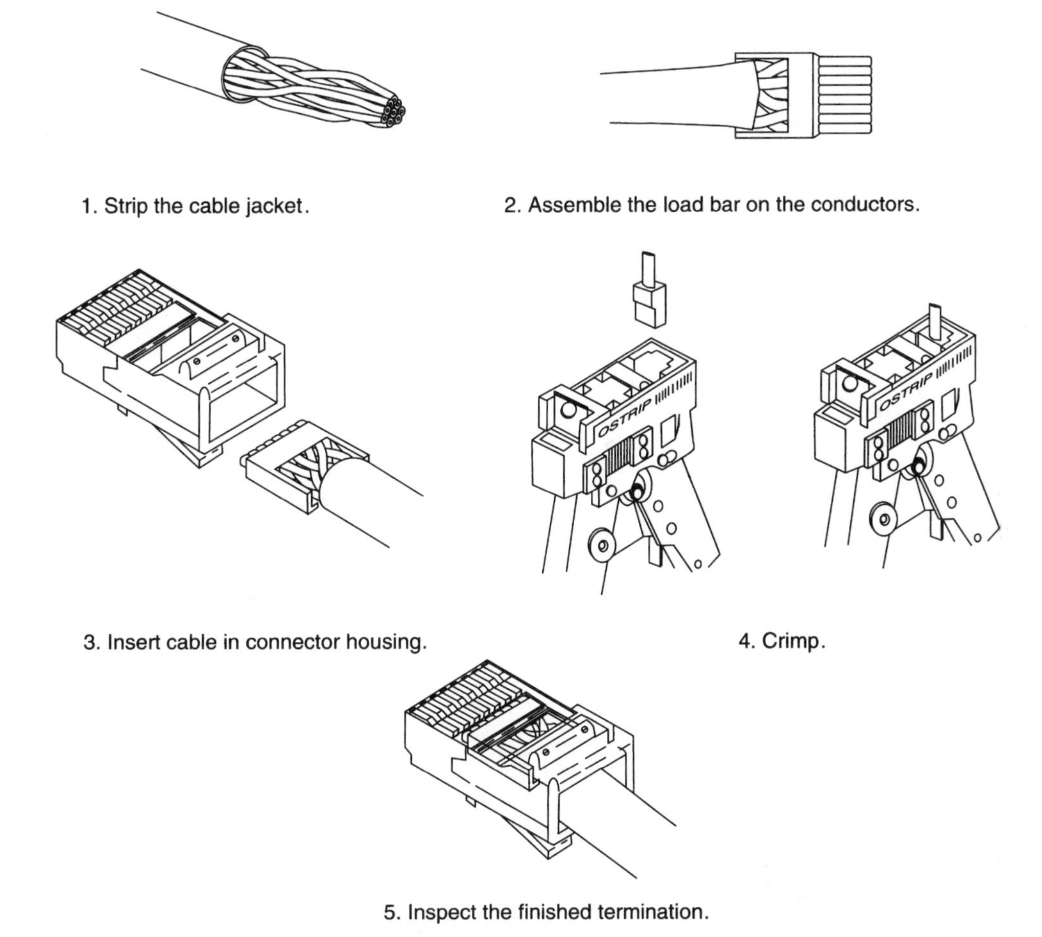

Figure 3–65 Terminating a modular plug.

- Make sure all components are rated to the category being installed.
- Use the same cabling scheme throughout the installation. That is, terminate each plug in the same wiring pattern. The preferred pattern is T568A or T568B. Don't, for example, mix T568A and T568B patterns. Using a mixed wiring pattern will only lead to future confusion and incompatibilities.
- Cross-over cables and connections, such as those required to connect two hubs should not be part of the premises cabling system. These should be treated as special cases outside of the system.

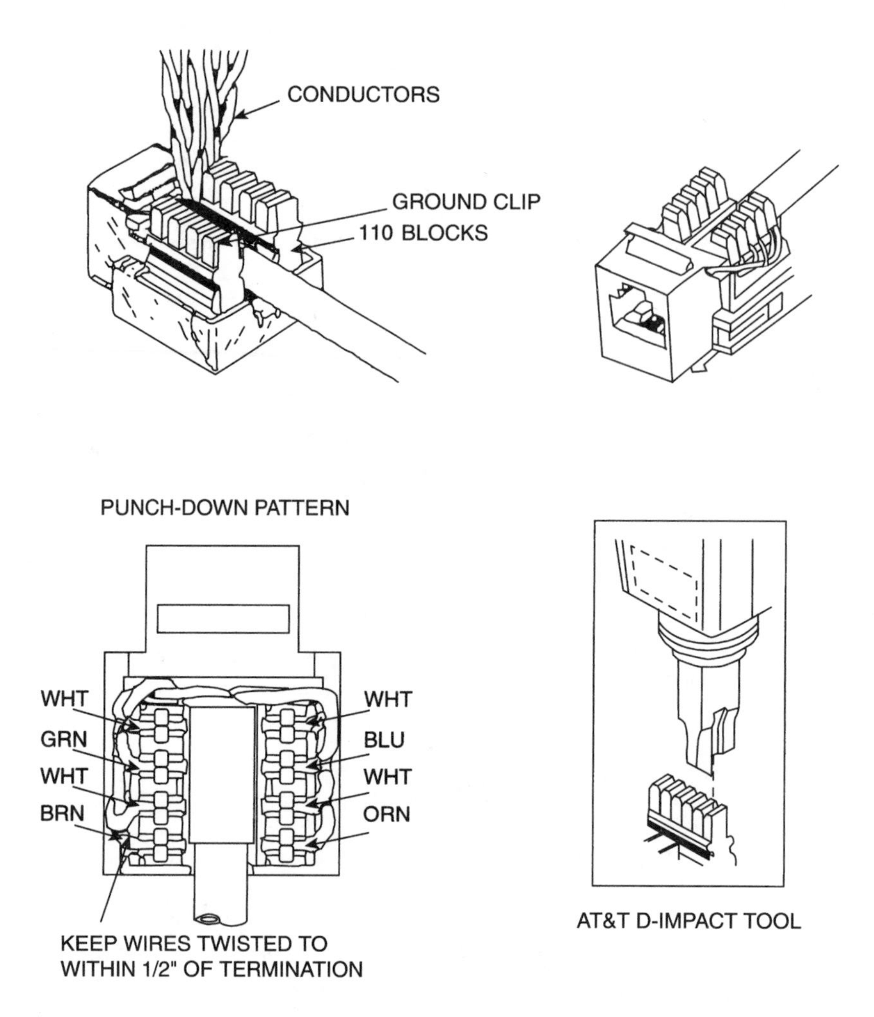

Figure 3–66 Terminating a 110-style modular jack.
(Courtesy of AMP Incorporated)

Trying to incorporate them into the cabling plant leads to long-term confusion and incompatibilities.

- Pay close attention to the position of each conductor within the connector. The pattern should be the same at both ends of the cable.
- While a connector can be reused, it's often faster and more cost-effective to cut off and discard an incorrectly applied connector.
- Make sure you have the proper connector for the cable. Some modular plugs are designed for stranded conductors, while others

accept solid conductors. Some plugs accept both stranded or solid conductors.

- Maintain proper cable-management practices with regard to tensile loads and bend radii. Neatness counts, but not at the expense of degraded cable.

Factory-terminated patch cables are preferred in the work area and at cross-connects. This is especially true for Category 5 cables, which are more prone to misapplication. Single-ended cables are available terminated at one end with a modular plug and unterminated at the other end. The unterminated end can be punched down at the cross-connect.

Terminating Fiber-Optic Connectors

Here are some useful things to remember about working with fiber:

- Use epoxyless connectors to speed application time. Eliminating epoxy cuts the number of consumables, cuts the potential for mess, and slashes installation time. The lack of epoxy also speeds polishing time since epoxy is harder to remove than glass. The connectors have proven themselves reliable and easy to use.
- Be careful with the fiber ends removed from the connectors. They can puncture the skin like a splinter and are hard to remove. Some installers use a piece of tape to collect them.
- Cleanliness is essential. Keep the fiber clean and dry when working with it. Dirt, films, and moisture are the enemies of fiber performance and reliability. Likewise, keep tools and supplies clean so they don't contaminate fibers.
- Limit the number of passes with the stripping blade. Once pass is recommended. Then clean the fiber with a lint-free pad soaked in pure (99 percent) isopropyl alcohol. Isopropyl alcohol of less than 99 percent purity can leave a film on the fiber.
- Maintain the proper bend radius. Most outlets and organizers have built-in fiber management features that will allow generous radii to be achieved.
- Leave slack at each interconnection, such as inside wall and floor outlets and organizers. At least 1 meter of fiber is recommended. This allows for future repairs or rearrangement of cabling.

Polarization of Fiber Optic Cables

TIA/EIA-568A recommends duplex SC connectors as the standard interface. The connectors are labeled *A* and *B*. This standard requires a cross-over at every coupling adapter. At a wall outlet, for example, the in-wall A-connector mates with the work area B cable, as shown in Figure 3–67. The cross-over simply means that one end of a specific fiber is labeled *A* and the other end is labeled *B*.

The SC system is polarized by a key on the connector and keyway on the adapter. It becomes important to maintain polarization throughout the installation so that transceivers on equipment are always properly connected. Adapters at each end of a cable run should be installed in the opposite manner from each other. In other words, if the adapter keyway is up at one end, it should be down at the other end.

For multifiber cables, pair even-numbered fibers with odd-numbered fibers. Fibers 1 and 2 form a pair, 3 and 4 form a second pair, and so forth.

This may seem overly complex, but it has a purpose. Consider two electronic devices, for example a hub and an NIC. For devices, the receiver port is A and the transmitter port is B. If the NIC connects directly to the hub, the transmitter port (B) of the NIC connects to the receiver port (A) of the hub. In other words, Port A at one end connects to Port B at the other end. What the 568 rules do is to maintain this A-to-B and B-to-A orientation at every adapter throughout the system. By thoughtfully ensuring polarization throughout the system, you save users from having to think about it later.

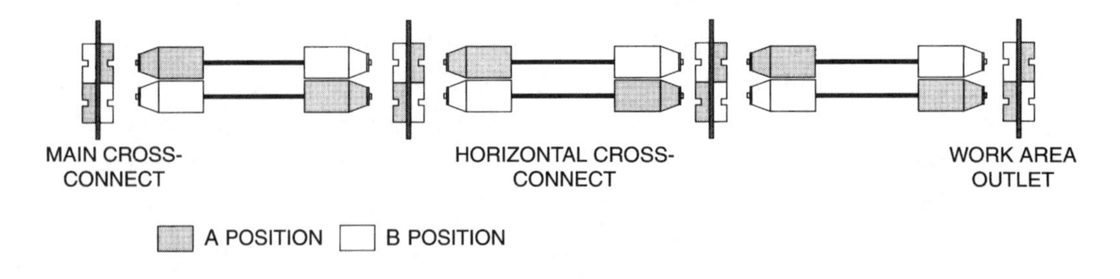

Figure 3–67 Polarization of duplex fiber optic cables.

Chapter 4

Cable Testing

KEY POINTS

- What is the difference between an acceptance test and a maintenance test?
- What is the purpose of electrically testing wire and cable?
- What type of test equipment is used to perform tests on wire and cable?
- What type of tests are performed on wire and cable?
- What specific safety precautions should be employed when testing wire and cable?

INTRODUCTION

No matter what its age, any piece of equipment is subject to damage. New wire and cable may have manufacturing defects or suffer shipping damage. Service-aged wire and cable are subject to normal deterioration caused by heat, vibration, chemicals, and electrical stress. Because of this, wire and cable should be tested when new, and thereafter periodically, to determine their serviceabilities.

This chapter describes the various facets of cable-testing equipment, procedures, and safety precautions. Most electrical cable tests are insulation tests; consequently, the material in this chapter focuses on insulation testing. Fiber optic cable tests, in contrast, are more often designed to test the quality of the path as well as the condition of splices and other intermediate equipment.

Note that the material in this chapter is *not* designed to create a "cable-test person." Rather, it stresses concepts and theories. For detailed training in cable testing, refer to one of the many training organizations offering hands-on experience.

TERMS

The following terms and phrases are used in this chapter.

Absorption Current: The current that flows as a result of mobile ionic charges in insulation. As the "capacitor" charges, it attracts ions in the insulation. The motion of these ions creates current flow.

Acceptance Test: A test performed on a piece of equipment when the equipment is new or newly installed.

Apparent Power: The total voltage multiplied by the total current in an alternating current circuit. The apparent power is measured in units of volt-amperes (va). Apparent power is equal to the vector sum of the true power and the reactive power.

Capacitive Current: The current that flows as a result of the charging of the cable "capacitor." For a cable, one "plate" is the center conductor, and the other is the shield or ground system.

Conduction Current: The actual leakage current that flows through the insulation. This current is analogous to the current flow through a conductor.

Continuity Test: A test that evaluates the condition of an electrical circuit. A circuit that is continuous has a relatively low resistance from one end to the other.

Cross-Talk: The energy, usually noise, that is coupled from one conductor to another in the same cable or between cables.

Ionic Charge: An electric charge caused by a surplus of positive or negative ions.

Ion: A particle, usually subatomic, that has a positive or negative electrical charge.

Maintenance Test: A test performed on a piece of equipment that has been in service for some time.

Megohmmeter: An instrument designed to measure very high resistances, typically over 1 million ohms. Megohmmeters usually feature relatively high test voltages of 100, 250, 500, 1,000, 2,500, 5,000, 10,000, or 15,000 volts and relatively high input impedances on the order of 500,000 ohms or more.

NEXT: Near-end cross-talk. NEXT is the cross-talk measured on the quiet line at the same end as the signal that is being induced into that line.

Overpotential Test: Overpotential testing, also called high potential testing, implies the use of voltage levels above the nominal or rated voltage of the cable.

Power Factor: The ratio of true power to apparent power. When multiplied by a 100 power factor, it is referred to as the "percent power factor."

Proof Test: A test that proves the capability of the system to withstand voltage or current. Most commonly used to describe an overpotential test wherein an insulation system is intentionally subjected to more voltage than designed. If the insulation withstands the proof test, it should be able to withstand the normal system voltage.

Reactive Power: The reactive portion of the apparent power. Reactive power is calculated by multiplying the applied voltage by that portion of the current that is 90 degrees out of phase with the voltage. Reactive power is measured in units of volt-amperes reactive (var).

Total Current: The sum of all component currents. This is the current that the meter measures. The total current starts at a relatively high value and gradually decreases to a final value equal to the conduction current. A current meter measuring total current flow starts high and gradually reduces to a final value. Inversely, a megohmmeter starts low and gradually increases to a final value.

True Power: The resistive portion of the apparent power. True power is calculated by multiplying the in-phase magnitudes of the voltage and the current. True power is measured in units of watts (w).

Velocity of Propagation: The speed at which an electromagnetic signal travels along a wire or fiber. In free space, light travels at 3×10^8 meters per second (300 million mps). In cable, it rarely exceeds 90 percent of that figure and usually is between 70 and 90 percent.

PURPOSE

Acceptance

Problems with new cable or newly installed cable are usually caused by either manufacturing defects or shipment damage. Before cable is placed into service, it should be checked to make certain that it is capable of performing correctly.

Insulation. Insulation tests are performed to ensure that the insulation has not been damaged or manufactured incorrectly. An insulation test evaluates the insulation for excessive leakage currents. Cracks, voids, or impurities in the insulation will cause excessive leakage current, which may result in eventual failure of the insulation. Acceptance tests will isolate these problems before the cable is allowed to go into service, thereby preventing an expensive, unplanned outage caused by cable failure.

Continuity. When power cable is installed, it must be terminated at each end and may be spliced at intermediate locations. These terminations and splices will exhibit high resistance if they are not properly made. The high resistance will cause excessive heat and ultimately result in insulation failure.

Control or communications cables may include a myriad of wires, each of which must be terminated at the proper location. If any wire is not terminated properly, the control system will fail to operate, causing down time, troubleshooting expense, or damaged equipment.

Power communications and control cable terminations and splices can be evaluated using continuity tests. These tests will spot installation problems before they have an opportunity to cause damage.

Circuit Certifications. Whether required by regulatory standards or simply common-sense types of tests, various procedures may be performed to evaluate or certify the operation of coaxial and twisted-pair communications cables. For the most part, these tests involve evaluating the ability of the cable to transmit signal with little or no distortion. These tests include:

- Wire map
- Length
- NEXT
- Attenuation
- Attenuation-to-cross-talk ratio
- Impedance
- Capacitance
- Loop resistance
- Noise

These tests are explained later.

Maintenance

As cable ages, various environmental and service conditions will cause it to deteriorate mechanically and electrically. Vibration can cause insulation layers to slip and crack, heat can cause terminations to loosen, and chemical agents can act on the insulation and reduce its resistance. Cable and wire, like all other pieces of equipment, have fixed lifetimes. How long they last depends on the environmental and service conditions they encounter.

Wire and cable must be evaluated periodically to determine serviceability. Proper evaluation of the test results can provide clues as to the life left in the cable.

Insulation. Vibration, heat, and chemical action all contribute to the aging process of insulation. Periodically, tests should be made to determine the condition of the insulation. By comparing test results at each maintenance interval, the continued serviceability of the insulation can be determined.

Continuity. Heat and vibration are the principal sources of high resistance problems. Thermal cycling will cause expansion and contraction of the termination, which will, in turn, cause the connection to loosen. Vibration aggravates the problem by shaking the connection and accelerating the loosening effects. Worse, as the connection loosens, the resistance and heat increase.

Periodic testing of connections will catch deteriorating connections before they have a chance to fail completely and cause major problems.

Circuit Certifications. Many of the tests described under the preceding "Acceptance" section also apply to communications circuits for maintenance. Because of the changes that may occur over time, whether from environmental or man-made changes, tests such as NEXT, attenuation, attenuation-to-cross-talk, impedance, capacitance, and noise may be performed.

ELECTRICAL INSULATION TESTERS AND TESTS

Various test sets may be used to test and evaluate both power and communications cable systems. The headings in the following sections include P, C, or PC designations to indicate that the equipment may be used on power cable (P), communications cable (C), or both (PC).

Megohmmeter (PC)

The megohmmeter is a direct voltage instrument designed to measure very high resistance values. To obtain sufficient sensitivity, megohmmeters employ relatively high voltages. Commonly used values include 100, 250, 500, 1,000, 2,500, 5,000, 10,000, and 15,000 volts. Even at these voltages, the high resistance of cable insulation allows only very small currents to flow. Consequently, the instrumentation employed in megohmmeters must be very sensitive.

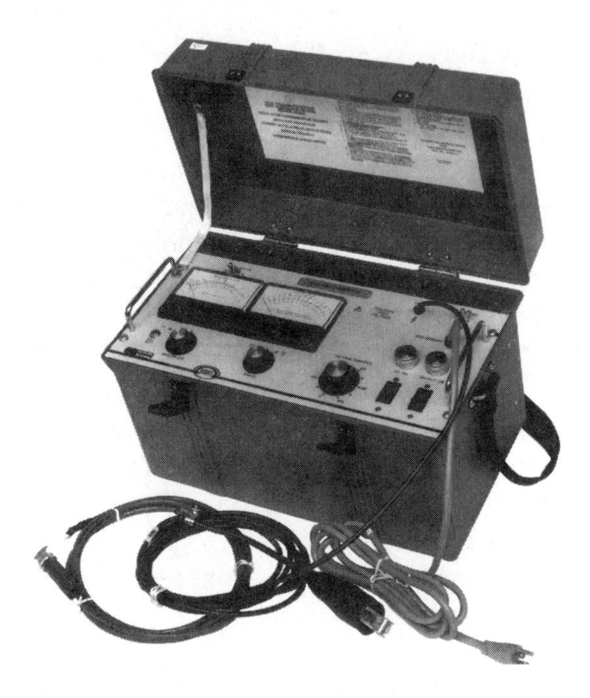

Figure 4–1 Typical megohmmeter.
(Courtesy of AVO, International)

Figure 4–1 shows a typical modern megohmmeter. Megohmmeters are available with varying degrees of sophistication. The key components of a megohmmeter are the meter(s), voltage control(s), and meter range switch(es).

A megohmmeter will typically have three leads, as shown in Figure 4–2. The line lead is the high-voltage lead that provides direct voltage to the test sample. The

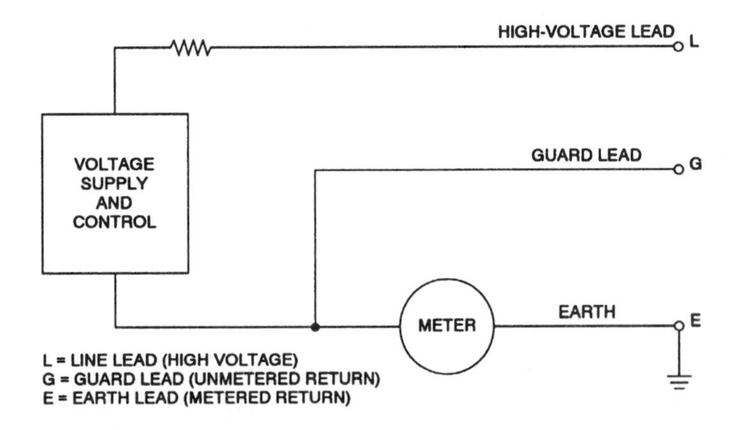

NOTE: SOME MEGOHMMETERS ARE CONSTRUCTED WITH THE METER AND GUARD LEAD IN THE HIGH-VOLTAGE CIRCUIT.

Figure 4–2 Simplified diagram of a megohmmeter.
(Courtesy of Cadick Corporation)

earth lead provides a return path for the current through the insulation specimen. The guard lead is also a return path lead, but it bypasses the meter.

The guard lead is employed when the tester does *not* wish to measure some part of the insulation. For example, if we wish to avoid measuring surface current on an insulator, the surface current will be returned to the instrument through the guard lead. Note that on some instruments the line lead is metered and the guard lead is at high potential.

Because insulation tests often take place on construction sites where electric service may be difficult to obtain or is poorly regulated, many megohmmeters are equipped with their own built-in, hand-cranked generators. Figure 4–3 is an example of a hand-cranked megohmmeter.

High-Potential Test Set (P)

High-potential, or overpotential, test sets differ from megohmmeters in several important ways:

- High-potential test sets have higher energy outputs. This translates to greater current capacity when performing a test.
- High-potential test sets are available in much higher voltage ranges than megohmmeters. High-potential test sets with outputs up to 500,000 volts are available.
- High-potential test set meters are *usually* micro-ammeters instead of megohmmeters. This means that test results are displayed in

Figure 4–3 Typical hand-cranked megohmmeter.
(Courtesy of AVO, International)

current values rather than megohms. While the resistance can still be calculated as the ratio of the voltage to the current, this is usually not required.

- Although direct voltage units are the most common, alternating voltage high-potential test sets are also available. Alternating voltage units have the advantage that they tend to stress insulation in the same manner as the system voltage. They have the disadvantage that qualitative measurements are difficult to make. Transformers and certain motor and generator circuits should never be tested with high direct voltage. Cable, however, may be tested with high direct voltage *as long as proper test procedures are observed.* See the section on "Types of Insulation Tests" later in this chapter.
- High-potential test sets are normally used to *intentionally* apply more voltage than the insulation rating. In this sense the high potential test becomes a "proof" test. If the insulation is able to withstand the overpotential, it will probably withstand the normal system voltage.

With the exceptions mentioned, high-potential test sets have most of the features of megohmmeters. In fact, high-potential test sets can be used to perform the same types of tests as megohmmeters. However, the megohmmeter is used for certain types of tests because of its smaller size and greater portability. Figure 4–4 shows a typical high-potential test set.

0.1 Hertz Test Set (P)

Even though manufacturers routinely apply overpotential to cable insulation for quality control, industry has long questioned the wisdom of intentionally applying excessive voltage to any piece of equipment. Several research studies completed long ago proved that properly applied direct voltage overpotential tests will not degrade or decrease the life expectancy of oil-impregnated, paper-insulated cable.

Some more recent studies, however, indicate that thermoplastic insulation may be damaged under the following conditions:

1. A direct voltage overpotential test applies more than twice the rated cable voltage.
2. The cable is not grounded for a sufficient time after the test is completed.
3. The power system voltage is applied at a worst case point on the waveform such that the applied voltage adds to the residual voltage in the cable.

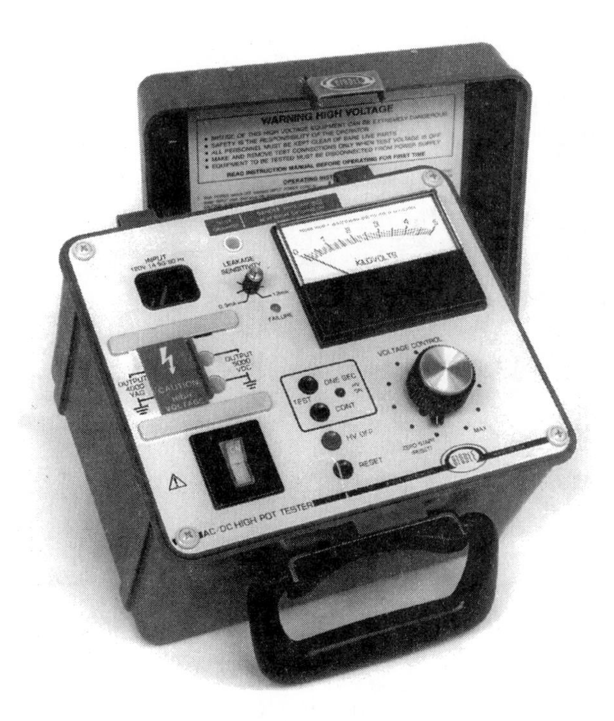

Figure 4–4 Typical high-potential test set.
(Courtesy of AVO, International)

The probability of all three of these events occurring during a test is relatively small, but it is not zero. Consequently, many testers, especially in Europe, have gone to the 0.1 hertz type of test.

A 0.1 hertz test set applies a 1/10 hertz, alternating voltage to the cable. This test has the advantages of DC testing, but can be used in such a way that little or no residual voltage is left on the cable. This test has become increasingly common in Europe, but it has not found widespread support in the United States. Figure 4–5 shows a typical 0.1 hertz test set.

Power Factor Test Set (P)

When an alternating voltage is applied to an insulation system, the resulting current flow is extremely reactive, leading the applied voltage by an angle that approaches 90°. If the insulation deteriorates or is damaged, the resistive current will increase while the capacitive current remains relatively constant.

Insulation can be evaluated based on the angle between the voltage and current, or power factor, which is the ratio of the real power (watts) consumed by the insulation to the total power (va) applied.

Figure 4–5 Very low-frequency test system.
(Courtesy of AVO, International)

A power factor test set, shown in Figure 4–6, is used to determine this angle. Power factor test sets are not often used for evaluation of cable because of the high capacitive current drawn by the cable.

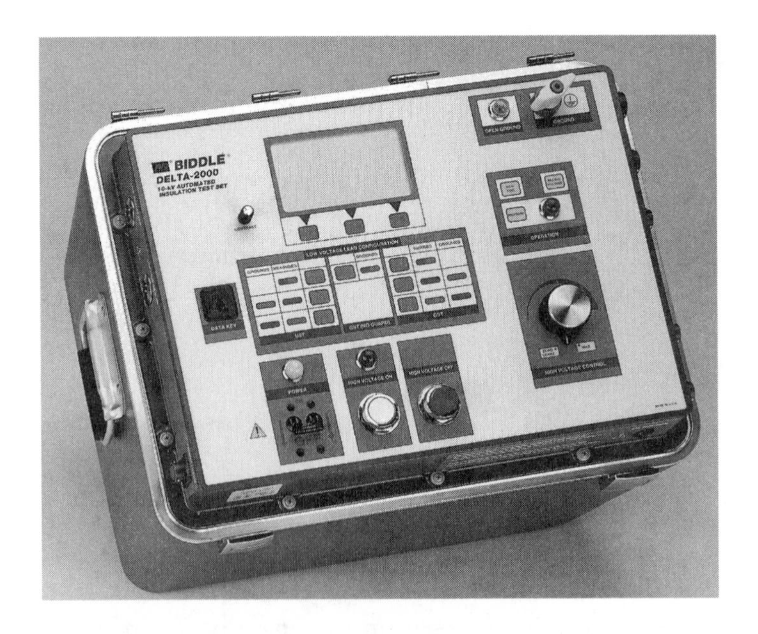

Figure 4–6 Typical insulation power factor test set.
(Courtesy of AVO, International)

ELECTRICAL CONTINUITY TESTS

Continuity tests fall into two basic categories: lead verification and resistance. These two tests, and the equipment required, are described in the following paragraphs.

Lead Verification

When multiconductor control cables are being terminated, the electrician must know where each wire end goes. A continuity test is often performed to aid in this operation. The procedure, which follows, is quite simple.

1. One end of the selected wire is grounded or otherwise connected to a known return conductor.
2. A continuity indicator, such as a multimeter or a simple battery-powered buzzer, is connected from the other end of the wire to ground.
3. A continuity indication shows that the selected wire is the correct one. This type of test is commonly performed during the construction phase of an electrical facility. The type of equipment used for this procedure is usually quite simple and often homemade. Figure 4–7 shows a typical buzzer test set used for this purpose.

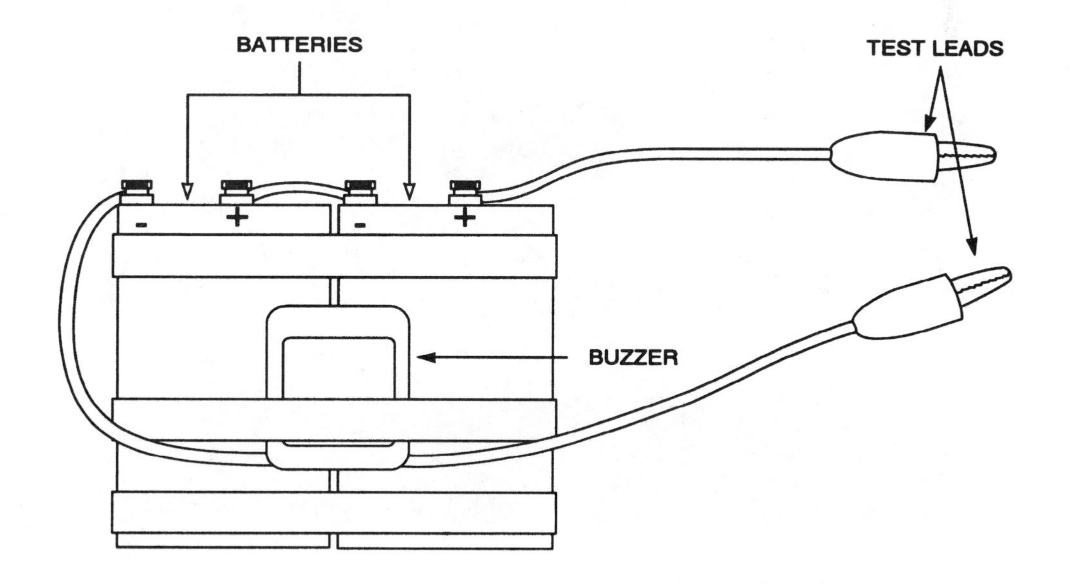

Figure 4–7 Buzzer test set for continuity testing.

The wire map test can be used to verify continuity and proper termination of multiconductor cables. The wire amp is discussed in more detail later in this chapter.

Resistance

Often the electrician will need to know the resistance of a connection, which is typically a few tens of micro-ohms. These values can be used to compare against previous readings to indicate a deterioration of the connection. The type of instrument used for such a test is called a micro-ohmmeter or a milli-ohmmeter, depending on the magnitude of the resistance present (see Figure 4–8).

To use this instrument, the current leads are first placed across the termination or joint that is to be evaluated (see Figure 4–9). The potential leads are then connected inside the current leads, as shown. Two lead sets are used so that the measurement is taken from the joint, as opposed to the test set ends of the current leads. In this way the measured resistance includes only the joint, not the leads or connections. Connections can be a significant part of the overall resistance. Current is applied from the test set and typically ranges from 10 to 400 amperes, with the resistance read directly from the meter.

This type of test is normally done as a maintenance or troubleshooting procedure.

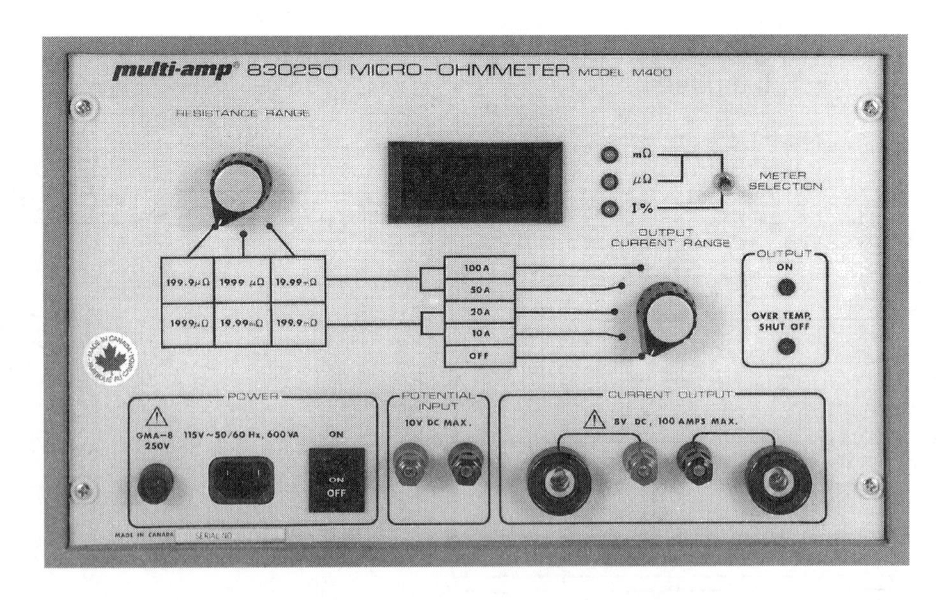

Figure 4–8 Typical micro-ohmmeter.
(Courtesy of AVO, International)

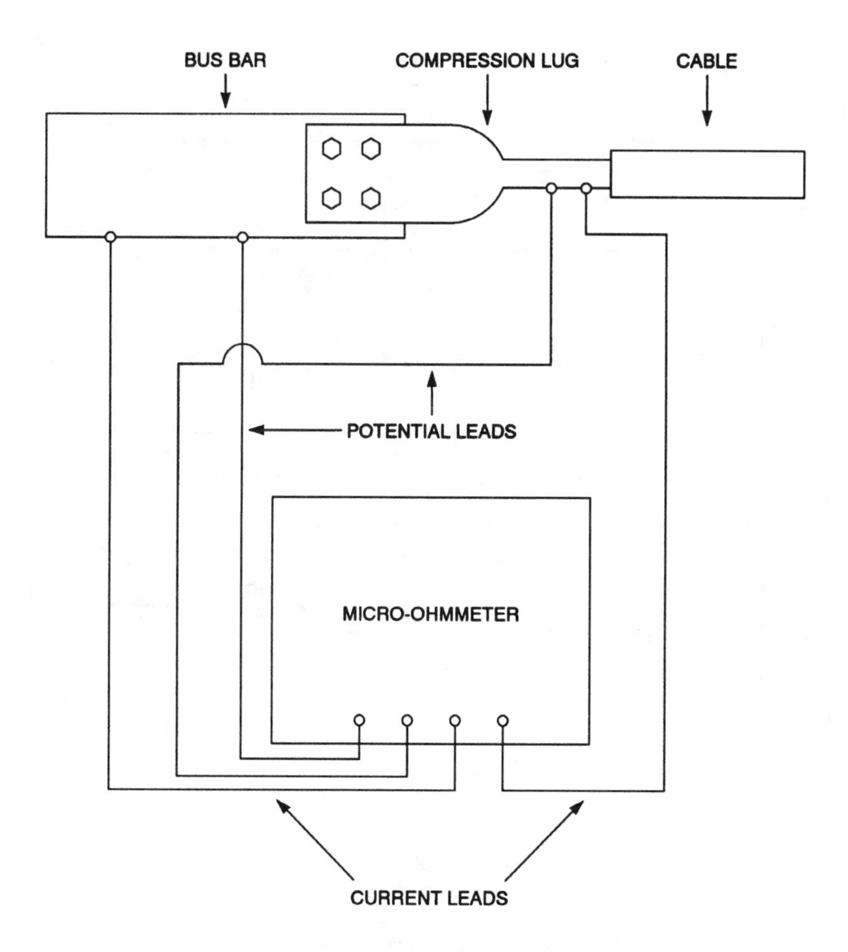

Figure 4–9 Using the micro-ohmmeter.
(Courtesy of Cadick Corporation)

INSULATION CURRENT

Wire and cable insulation presents a capacitive circuit for the passage of electrical current; see Chapter 1 for more information. Figure 4–10 shows a graphical representation of the current that will flow through insulation when a direct voltage is suddenly applied to it. The different types of current flows—conduction, capacitive, absorption, and total—are defined in the "Terms" section at the beginning of this chapter.

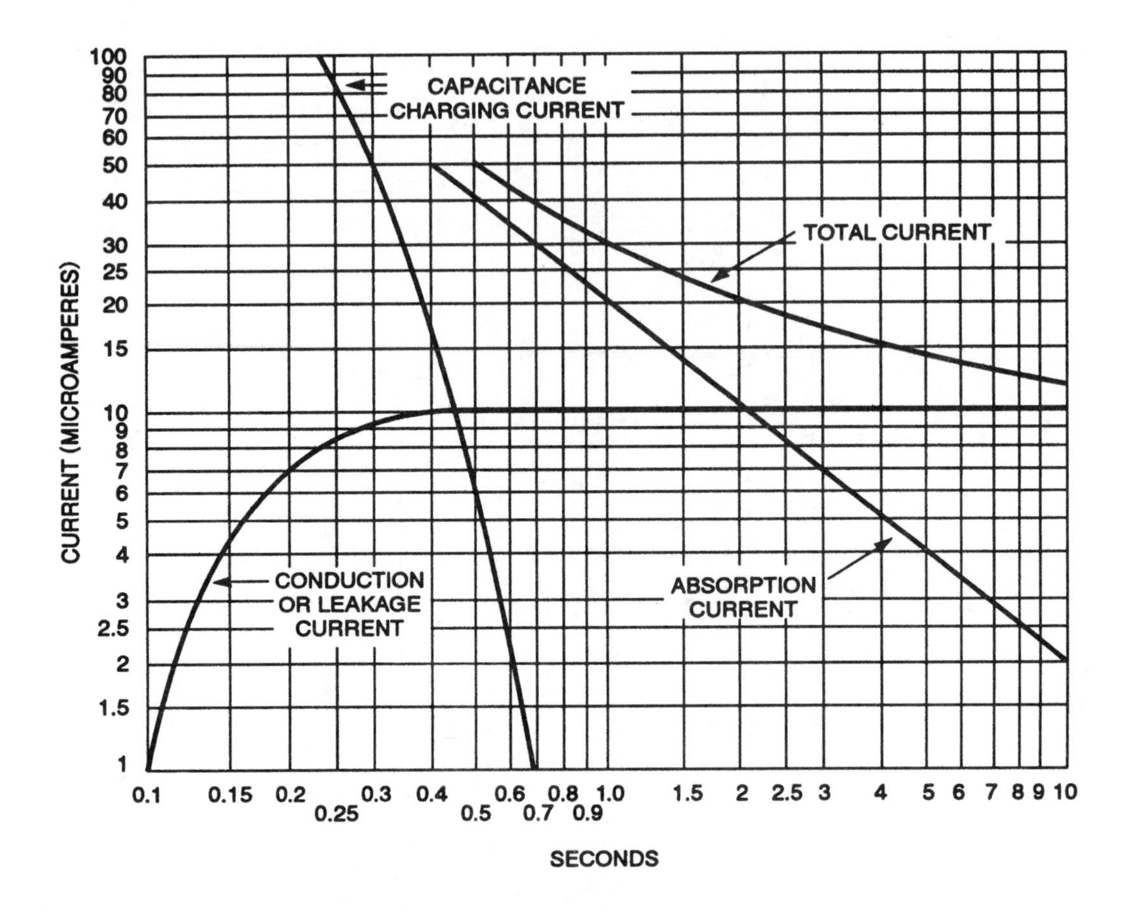

Figure 4–10 Curves showing components of current measured during DC testing of insulation.

TYPES OF INSULATION TESTS

CAUTION:

THE TEST PROCEDURES GIVEN IN THE FOLLOWING SECTIONS ARE FOR REFERENCE ONLY! PERSONS UNFAMILIAR WITH THE OPERATION AND SAFETY HAZARDS ASSOCIATED WITH THIS TYPE OF TESTING SHOULD NOT PERFORM ANY OF THESE TESTS WITHOUT APPROPRIATE SUPERVISION. THESE TEST PROCEDURES ARE NOT INTENDED TO BE USED FOR AN ACTUAL TEST. USE *ALL* APPROPRIATE SAFETY PROCEDURES INCLUDING, BUT NOT LIMITED TO, THOSE LISTED AT THE END OF THIS CHAPTER, MANUFACTURER'S RECOMMENDATIONS, ELECTRICAL CONTRACTOR'S PRACTICES, AND CUSTOMER'S/OWNER'S SPECIFICATIONS.

Insulation Resistance

Figure 4–11 shows a megohmmeter connected to measure the insulation resistance of one conductor of a three-conductor cable. This particular configuration will give the resistance of the conductor with respect to ground. The test procedure is as follows:

1. FOLLOW ALL APPROPRIATE SAFETY STEPS FOR THIS TEST.
2. Select a proper voltage range for the test. Low-voltage cable may be tested at 500 volts or 1,000 volts. Refer to the manufacturer's recommendations for medium-voltage cables.
3. Ground all conductors that will not be part of the measurement.
4. Connect the earth lead of the megohmmeter to the conductor being tested.
5. Connect the line lead of the megohmmeter to the conductor being tested.

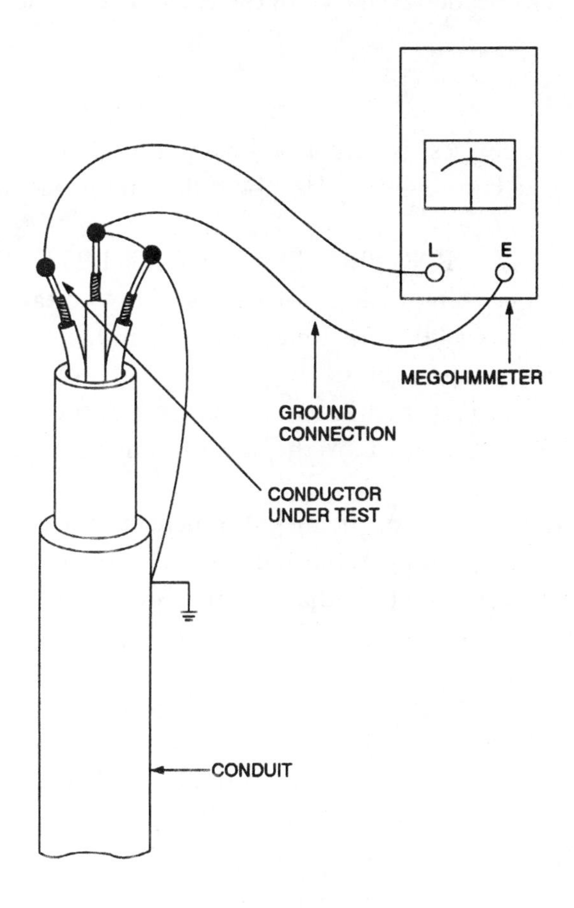

Figure 4–11 Cable test.

6. Energize the megohmmeter at the voltage selected in Step 2.
7. Wait 1 minute, taking note that the indicated insulation resistance is gradually rising. This rise occurs because the capacitive and absorption currents in the cable are decreasing.
8. Read and record the insulation resistance from the megohmmeter scale after 1 minute. One minute is chosen as a reasonable time to allow the current to settle.
9. Locate an appropriate temperature correction table for the insulation being tested, and correct the measured resistance for 20° Celsius.
10. Record all measured and corrected values and compare them to previous readings and local standards.
11. If the megohmmeter switch has a drain or ground position on its selector switch, turn it to this position long enough to drain any static charge from the cable (from 1 minute to ½ hour depending on the cable size).
12. Reconnect as needed to measure the resistance of the other conductors.

Polarization Index

A polarization index test is performed using the same setup as for the insulation resistance test (see Figure 4–11). Use the following procedure:

1. FOLLOW ALL APPROPRIATE SAFETY STEPS FOR THIS TEST.
2. Select a proper voltage range for the test. Low-voltage cable may be tested at 500 volts or 1,000 volts. Refer to the manufacturer's recommendations for medium-voltage cables.
3. Ground all conductors that will not be part of the measurement.
4. Connect the earth lead of the megohmmeter to the conductor being tested.
5. Connect the line lead of the megohmmeter to the conductor being tested.
6. Energize the megohmmeter at the voltage selected in Step 2.
7. Wait 1 minute, taking note that the indicated insulation resistance is gradually rising. This rise occurs because the capacitive and absorption currents in the cable are decreasing.
8. Read and record the insulation resistance from the megohmmeter scale after 1 minute.
9. Continue the test an additional 9 minutes for a total of 10 minutes. Take and record another reading after 10 minutes.
10. Divide the 10-minute reading by the 1-minute reading and compare to local standards. Table 4–1 is a table of generally acceptable values for the polarization index.

Table 4–1 Dielectric absorption ratios.

Insulation Condition	60/30-Second Ratio	10/1-Minute Ratio Polarization Index
Dangerous	—	Less than 1
Questionable	1.0 to 1.25	1.0 to 2
Good	1.4 to 1.6	2 to 40
Excellent	Above 1.6	Above 4

11. If the megohmmeter switch has a drain or ground position on its selector switch, turn it to this position long enough to drain any static charge from the cable (from 1 minute to ½ hour depending on the cable size).
12. Reconnect as needed to measure the polarization index of the other conductors.

Overpotential Tests

A direct-voltage overpotential test may be performed on medium-voltage cable for maintenance or acceptance testing purposes. Overpotential testing is not usually performed on low-voltage unshielded cables. Figure 4–12 shows an AVO, International Model 220110 test set connected to perform an overpotential test on a cable. The following explains each of the connections shown in that figure.

The AVO, International Model 220110 DC test set consists of two units. Cable *C1* is the connecting cable between the control unit and the high-voltage transformer unit. Connection *A2* is the line (high-voltage) connection, and *A1* is the guard connection. The guard connection is normally used in these tests because the extremely high-voltage levels will cause surface leakage currents to flow that will interfere with the test results. Cable C2 contains the 120-volt line cord, the safety foot switch, and the ground connection. Notice that the untested cables are grounded and jumpered together.

The overpotential test is performed by applying voltage to the cable in five to ten equal steps up to a maximum value. At each step, the current is read after the time varying components have settled out; this interval is usually from 1 to 5 minutes. When the maximum value is reached, the voltage is left on for 15 minutes, with current readings taken every minute. The following procedure may be used:

1. FOLLOW ALL APPROPRIATE SAFETY STEPS FOR THIS TEST.
2. Select a proper maximum voltage for the test. Refer to Table 4–2 or local engineering standards to determine this voltage. Manufacturer's recommendations will always take precedence over other sources of voltage information. Note that Table 4–2 applies only to new cable. Do not use values for old service-aged cable.

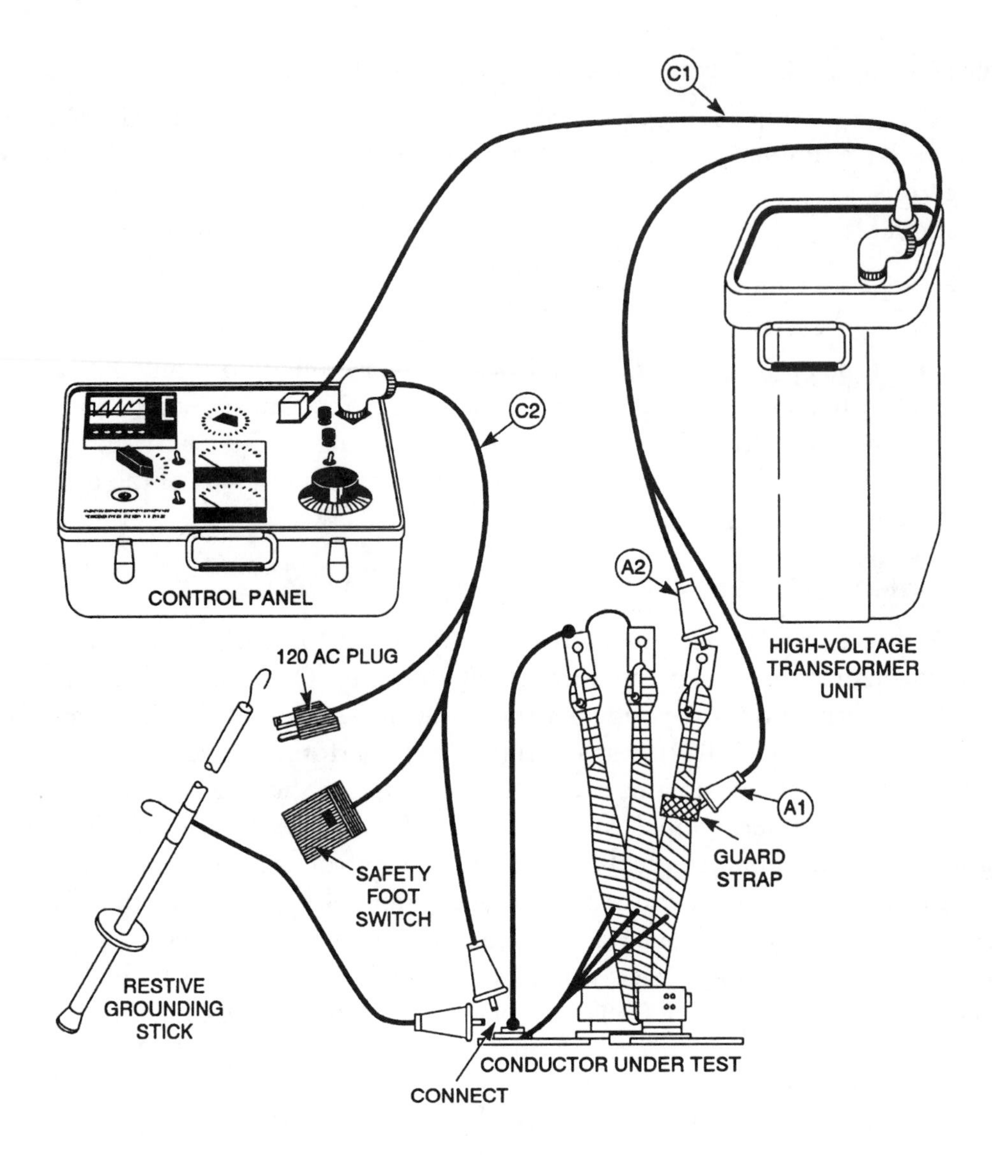

Figure 4–12 Typical test connection using the 220110 DC Dielectric Test Set (110 kV). *(Courtesy of AVO, International)*

3. Ground all conductors that will not be part of the measurement.
4. Connect the earth lead of the test set to the ground.
5. Connect the line lead of the test set to the conductor being tested.
6. Connect the guard lead of the test set to the insulation of the conductor being tested. This connection should be made using a connection strap placed 6 inches to 1 foot away from the terminal connections.

Table 4–2 Maximum voltage for acceptance testing of new cable. *(Courtesy of NETA Acceptance Testing Specification, 1996, International Electrical Testing Association, PO Box 687, Morrison, CO 80465)*

Insulation Type	Insulation Level (Percent)	Rated Cable Voltage (kV)	Test Voltage (kV dc)
Elastomeric butyl and oil base	100	5	25
	100	15	55
	100	25	80
	133	5	25
	133	15	65
Elastomeric: EPR ethylene-propylene rubber	100	5	25
	100	15	55
	100	25	80
	100	35	100
	133	5	25
	133	15	65
	133	25	100
XL, XLP polyethylene	100	5	25
	100	15	55
	100	25	80
	100	35	100
	133	5	25
	133	15	65
	133	25	100

NOTE: For acceptance testing new cable only. Do not use these voltages for service-aged cable. Derived from ANSI/IEEE Std. 141–1986 Table 82.

7. Determine the number of voltage steps to be used in the test. Divide the maximum test voltage by the number of steps and round to the nearest kilovolt. For example, if the maximum test voltage is 40 kV and the number of steps is 8, the step voltage will be 40 kV/8 = 5 kV.
8. Make certain that all personnel are clear at *both* ends of the cable.
9. Energize the test set and *gradually* increase the voltage to the first step level (5 kV from the example in Step 7). Note that the current rises initially and gradually decreases. Wait 1 to 5 minutes to read and record the leakage current.
10. Gradually increase the voltage to the next step level (10 kV in our example). When the next step voltage is reached, wait the same amount of time as in Step 9. Read and record the leakage current.

11. Repeat Step 10 for each of the voltage steps up to and including the maximum test voltage. *At any step, if the current continues to increase after the step voltage is reached, terminate the test immediately. The cable has failed.*
12. Leave the final voltage on the cable for 15 minutes, reading and recording the leakage current every minute. The current should continue to decrease during this period. All readings should be entered into a data table similar to the one shown in Table 4–3.
13. Gradually decrease the test voltage to 0. When the voltmeter on the test set indicates 0 volts, put a ground wire on the cable and leave it for a minimum of 30 minutes.

CAUTION:

THE CABLE WILL CONTINUE TO DRAIN CHARGE FOR THE ENTIRE PERIOD THAT THE GROUND WIRE IS ATTACHED. DO NOT REMOVE THE GROUND WIRE OR TOUCH THE CABLE WITHOUT WEARING PROPER INSULATING GLOVES AT ANY TIME!

14. Plot the values of voltage versus current and current versus time on a graph similar to the one shown in Figure 4–13.

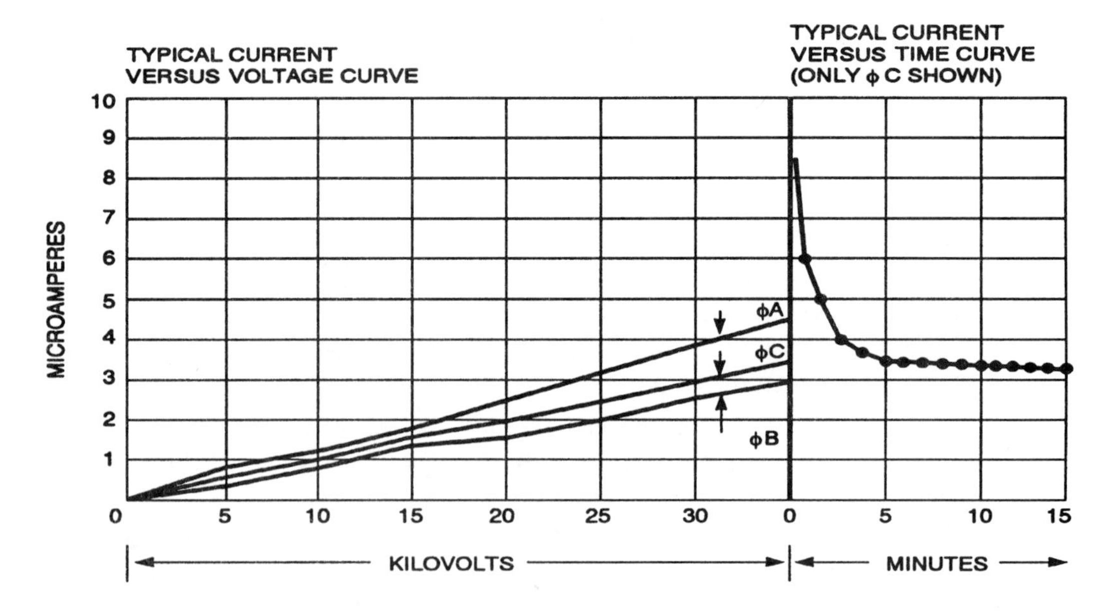

Figure 4–13 Typical graph showing current versus voltage curves and current versus time curves.

Table 4–3 Dielectric strength test record.

Cable	Location
Type	Rating
Date	Weather
Remarks	

Af		Bf		Cf	
KV	mA	KV	mA	KV	mA
5	0.8	5	0.4	5	0.6
10	1.2	10	0.8	10	1
15	1.8	15	1.25	15	1.4
20	2.6	20	1.6	20	2
25	3.2	25	2.15	25	2.6
30	3.95	30	2.6	30	3
35	4.6	35	3.15	35	3.6
1 MIN		1 MIN		1 MIN	6.0
2 MIN		2 MIN		2 MIN	4.75
3 MIN		3 MIN		3 MIN	4.25
4 MIN		4 MIN		4 MIN	3.80
5 MIN		5 MIN		5 MIN	3.60
6 MIN		6 MIN		6 MIN	3.58
7 MIN		7 MIN		7 MIN	3.57
8 MIN		8 MIN		8 MIN	3.55
9 MIN		9 MIN		9 MIN	3.50
10 MIN		10 MIN		10 MIN	3.48
11 MIN		11 MIN		11 MIN	3.44
12 MIN		12 MIN		12 MIN	3.42
13 MIN		13 MIN		13 MIN	3.40
14 MIN		14 MIN		14 MIN	3.40
15 MIN		15 MIN		15 MIN	3.40

0.1 Hertz Testing

As explained previously, the 0.1 hertz test is becoming increasingly popular in some parts of the world. It offers the advantage of lighter equipment than 60 hertz alternating voltage, yet it stresses the insulation as an alternating voltage would.

Direct voltage testing can cause very large charge densities to build in the area of insulation imperfections. These charge densities may take hours to drain, even with a direct short across the cable insulation. If the cable is re-energized in such a way that the applied alternating voltage adds to the test charges, the cable may fail. This is not a problem with 0.1 hertz testing, because the voltage alternates; therefore, no charge builds, and the cable is not subjected to very high voltages when it is re-energized.

These problems described are more of a concern during maintenance testing; new cable should not exhibit problems of this type.

Power Factor Testing

The principles of power factor testing were explained earlier in this chapter. When a power factor test is performed on insulation, several quantities may be measured as follows.

- The watts consumed by the insulation may be measured. This is a measure of the absolute value of the leakage current; high values may indicate a general failure or deterioration of the insulation.
- The power factor (watts divided by va) may be measured. This value can be compared to readings taken previously and from similar equipment. Greater than normal values usually indicate a problem.
- Power factor at different voltage levels can be measured. If the power factor increases as the voltage increases, the insulation may be contaminated.

Measurement of power factor on long cable runs is difficult because of the extremely high capacitive currents. These currents can be offset by the use of a parallel, resonating inductor. The inductor cancels out some of the capacitance of the cable. The readings are then taken as usual and compared to similar readings.

For the most part, power factor tests are usually not performed on cables.

ELECTRICAL COMMUNICATIONS CABLE MAINTENANCE AND CERTIFICATION PROCEDURES

Figure 4–14 shows a cable tester (on the left) and a signal injector. Such instruments can be used to test and evaluate copper communications cables. The signal injector is used to inject voltages of known magnitude and duration, while the tester is used to measure and evaluate the results. The tests that are performed can include or exclude the patch cables that connect the equipment to the main network cable. A link test includes the patch cables, while a channel test does not. Figure 4–15 shows the two basic circuit configurations.

The types of tests that can be performed and the expected results are described in the following sections.

Wire Maps

A wire map test is a somewhat more sophisticated form of the continuity tests described earlier in this chapter. In this test, the cable tester is connected to one end

Figure 4–14 Cable tester and signal injector. *(Courtesy of Microtest)*

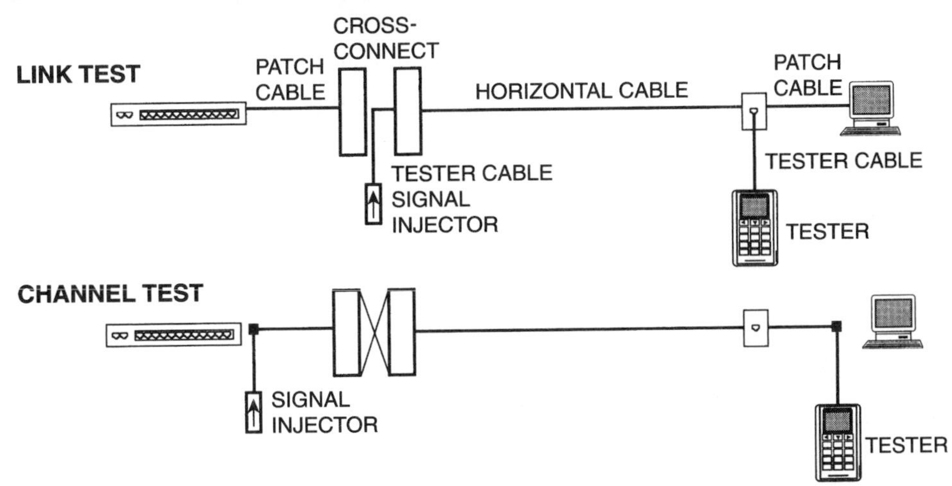

Figure 4–15 General setup for link and channel testing.

of the cable and the signal injector is connected to the other. Usually the instruments are connected to all pairs at once.

The signal injector injects signals that are unique to each pair, and the tester receives and evaluates the signals, ensuring that the correct signal is found on the correct pair and with the correct polarity. Figure 4–16 shows some of the possible incorrect connections a wire map test may uncover.

Cable Length

In addition to checking wire connections, the cable tester can determine cable length in a relatively straightforward procedure. The tester generates a pulse, which travels down the cable. When the pulse reaches the remote end of the cable, a reflection bounces back to the tester. The tester times the round trip and calculates the cable length using a simple time, rate, and distance formula:

$$L = \frac{v \times t}{2} \tag{1}$$

Where:

L = the length of the cable
v = the velocity of propagation in the cable
t = the time of the round trip made by the pulse

Note that the denominator is 2. This is because the pulse must make the trip both ways. Another critical concern is the velocity of propagation in the cable. Many manufacturers publish the velocities of propagation for their products. If manufacturer's figures are not available, the velocity of propagation can be calculated by measuring the round trim time *(t)* on a known length of cable and using it to calculate v.

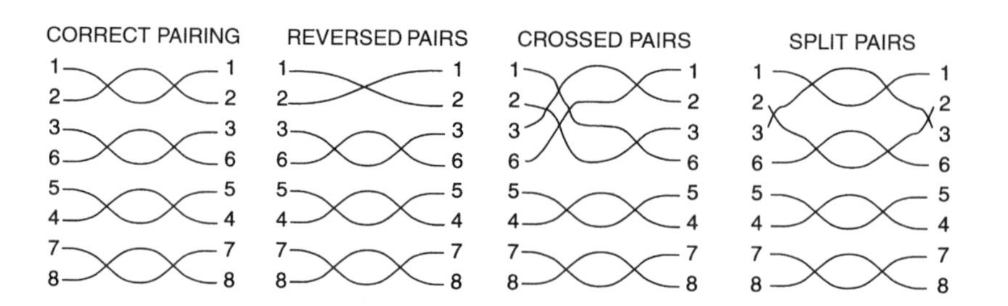

Figure 4–16 Continuity test for reversed, crossed, and split pairs.

As the next chapter describes, this reflection technique can also be used to find problems in the cable.

NEXT Loss

In a NEXT loss test, the tester injects a series of frequencies onto the transmit pair and measures the energy that is coupled into the receive pair. Ideally, the amount of coupled energy is 0. The tester will couple frequencies that range from 700 to 100 MHz for Category 5 cable or coaxial cable. Category 4 cable is tested only to 20 MHz; other cables are tested at lower maximum values.

The receiver used in the tester is a narrow band receiver to eliminate its response to noise or signals other than the test frequencies.

Attenuation

As mentioned earlier, signals are attenuated or reduced in magnitude as they travel down a cable. To ensure that the attenuation of the cable is at rated value or better, a tester can inject a signal of known amplitude into one end of the cable and measure it at the other. The decrease in signal is the attenuation. Attenuation tests are also made over the entire range of the signal.

Capacitance

Capacitance can be measured using various techniques. Power cable test sets, described earlier in this chapter, apply a rated frequency to the cable and measure the resulting current magnitude and phase angle.

Communications cable testers normally test capacitance by injecting a known amount of charge and measuring the rate of increase or decay of the charge. Capacitance can also be checked by measuring the voltage after a known amount of charge has been injected. The capacitance can be calculated using the following equation:

$$C = \frac{Q}{V} \tag{2}$$

Where:

C = the capacitance in farads
Q = the charge in coulombs
V = the voltage in volts

Note that farads is too high a unit of measure for this test. Normally, the results of this type of test would be in microfarads (μf – 10^{-6}) or picofarads (pf – 10^{-12}).

DC Loop Resistance

DC loop resistance is calculated by measuring the DC resistance of the cable going both ways. That is, one pair is connected and the round-trip resistance is measured from the other end. Coaxial cable is checked by connecting the shield to the center conductor and then measuring the round-trip DC resistance.

Impedance

Cable characteristic impedance was defined and discussed in Chapter 2. Field measurements using a time domain reflectometer (TDR) can evaluate cable characteristic impedance to ensure that the impedance is within allowed tolerance. Accurate impedance is particularly important to ensure that there are no reflections in the cable with their attendant increased attenuation. Most coaxial cables used in communications networks are 50 Ω or 75 Ω (see Chapter 2). Category 3, 4, and 5 cables have nominal impedances of 100 $\Omega \pm 15$ percent.

Determining Pass/Fail

Pass/fail is determined by comparing the test results to published standards, such as the Telecommunications Industry Association (TIA) or the International Standards Organization (ISO). To conduct such a comparison, the tester first generates a test report like the one in Figure 4–17. He or she then compares the results to information like that in Table 4–4. Any test results that deviate by more than the allowed percentage indicate problems that should be investigated further.

TESTING FIBER OPTIC CABLE

Fiber optic cable is tested using an optical time domain reflectometer (OTDR). This instrument and its testing are described in Chapter 5.

Godfrey Cyberwidgets
PENTASCANNER CABLE CERTIFICATION REPORT
CAT5 Link Autotest

Circuit ID:	1	Date:	20 March 95
Test Result:	PASS	Cable Type:	Cat 5 UTP
Owner:	PentaScanner	Gauge:	
Serial Number:	38S95BB_KLW	Manufacturer:	Belden
Inj. Ser. Num:	38N94BB_KLW	Connector:	Mod Plug
SW Version:	V03.10	User:	DJS
Building:	Engineering	Floor:	Second
Closet:	2A-N		
Rack:	1	Hub:	10B-T/1
Slot:		Port:	13

Test	Expected Results	Actual Test Results
Wire Map	Near: 12345678 Far: 12345678	Near: 12345678S Far: 12345678S

Test	Unit	Expected Results	Pr 12	Pr 36	Pr 45	Pr 78
Length	ft	10 - 328	95	95	97	95
Impedance	ohms	80 - 125	112	115	111	115
Resistance	ohms	0.0 - 18.8	4.8	5.2	4.9	5.1
Capacitance	pF	10 - 5600	1206	1189	1286	1219
Attenuation	dB		4.9	4.2	4.9	4.6
@Freq	MHz		100.0	100.0	100.0	100.0
Limit	dB		24.0	24.0	24.0	24.0

PENTA Pair Combinations		12/36	12/45	12/78	36/45	36/78	45/78
NEXT Loss	dB	39.3	40.0	37.5	32.8	37.6	39.5
Freq(0.7-100.0)	MHz	94.3	97.3	87.9	96.7	99.9	96.1
Limit: Cat 5 formula	dB	27.5	27.3	28.0	27.3	27.1	27.4
Active ACR	dB	38.3	45.6	34.9	32.1	33.0	36.1
Frequency	MHz	100.0	62.5	100.0	100.0	100.0	100.0
Limit: Derived	dB	3.1	12.1	3.1	3.1	3.1	3.1

INJ Pair Combinations		12/36	12/45	12/78	36/45	36/78	45/78
NEXT Loss	dB	37.0	38.6	34.2	31.7	38.9	43.6
Freq(0.7-100.0)	MHz	99.9	97.5	88.1	93.3	99.9	96.1
Limit: Cat 5 formula	dB	27.1	27.3	28.0	27.6	27.1	27.4
Active ACR	dB	32.1	38.8	36.4	31.9	34.3	38.9
Frequency	MHz	100.0	100.0	100.0	100.0	100.0	100.0
Limit: Derived	dB	3.1	3.1	3.1	3.1	3.1	3.1

Signature: ______________________ Date: __________

Figure 4–17 Printout of test results.
(Courtesy of Microtest)

Table 4–4 UTP cable specification ranges (TIA/EIA–568A and TSB–67).

Characteristic	Frequency	Category 3	Category 4	Category 5
NEXT Loss	0.772	43	58	64
	1	41	56	62
	4	32	47	53
	8	27	42	48
	10	26	41	47
	16	23	38	44
	20	—	36	42
	25	—	—	41
	31.25	—	—	39
	62.5	—	—	35
	100	—	—	32
Attenuation	0.772	2.2	1.9	1.8
	1	2.6	2.2	2.0
	4	5.6	4.3	4.1
	8	8.5	6.2	5.8
	10	9.7	6.9	6.5
	16	13.1	8.9	8.2
	20	—	10.0	9.3
	25	—	—	10.4
	31.25	—	—	11.7
	62.5	—	—	17.0
	100	—	—	22.0
Impedance		100 ohms ± 15%		
Structural Return Loss	1	12	21	23
	4	12	21	23
	10	12	21	23
	16	10	18	23
	20	—	19	23
	31.25	—	—	20.9
	62.5	—	—	18
	100	—	—	16
Mutual Capacitance (pF)	1 kHz	6.6	5.6	5.6

SAFETY PRECAUTIONS

NOTE:

The following safety precautions are typical and represent generally accepted safety practices. These precautions may not cover every installation or every variation. The worker should always refer to local safety standards, which always take precedence.

Area Security

The immediate area at each end of the cable to be tested should be secured by constructing barricades and danger signs should be posted. A guard should be stationed at the remote end of the cable being tested and appropriate communications should be established.

Equipment Isolation

All connected equipment should be disconnected from the cable before the test begins. If this is not possible, the test voltages should be reduced to levels that are acceptable for the connected equipment.

Circuit and Cable Isolation

1. Identify a lead person for the job and assign duties.
2. Isolate, lock, tag, and test the equipment that is in the testing area.
3. Post danger signs to identify the testing area.
4. Notify all personnel of the potentially hazardous conditions.
5. Inspect all rubber protection equipment before it is placed in service.

Safety Precautions Before Testing

1. Identify the tests to be performed and the test equipment to be used.
2. Identify the proper protective equipment for each test.
3. Review prints and clearances for each test.
4. Put rubber mats or sleeves on exposed conductors.
5. Be sure all switches in the circuits are open.
6. Remove jumpers, potential transformers, lightning arresters, and other dangerous materials.
7. Be sure the cable under test is de-energized before touching it by testing the circuit with the proper voltage test instruments and performing the proper lockout/tagout procedures.
8. Ground the cable after the circuit is de-energized.

Safety Precautions During Testing

1. Position personnel at both ends of the cable.
2. Apply ionization control (protection) where required on cable ends. Ionization control may be in the form of corona balls or insulating covers for the exposed cable ends.

3. Always wear appropriate rubber insulating clothing, such as gloves and sleeves.

Safety Precautions After Testing

1. Ground the tested conductor for a period at least as long as the test voltage was present. Longer grounding times are preferred.
2. Measure the conductor voltage after removing the ground and before touching the conductor. Wear rubber insulating equipment and always treat the exposed conductor as though it is still energized.

Safety Grounding

In addition to grounding the cable for the purpose of draining the charge, cable should be routinely safety grounded at all times while personnel are working in the area. The following steps are recommended:

1. Always wear rubber gloves with appropriate leather protectors.
2. Verify that the circuit to be tested is de-energized. This is accomplished with the use of an appropriate voltage measuring tool. When measuring for a de-energized circuit, always check the instrument on a known energized source before and after measuring the circuit to be tested.
3. After the circuit is verified de-energized, connect an approved safety ground to the ground bus.
4. Connect the other end of the safety ground to the conductor. Make these connections with "hot sticks" when possible.
5. After each of the conductors is properly grounded, it can be disconnected and separated for testing.
6. The conductors should remain grounded at all times, except when an individual conductor is under test. The safety grounds should first be removed from the cable, then from the ground.

Chapter 5

Cable Fault Location

KEY POINTS

- What is the purpose of fault location?
- What are the common types of fault location equipment?
- What are the common methods of fault location?

INTRODUCTION

Occasionally, even newly installed cable will fail. When this happens, the cable may be completely replaced or repaired, depending on economics. If the decision is to repair the cable, the failed section of the cable must first be located.

This chapter describes the two most commonly employed methods of cable fault location and the equipment that is used.

TERMS

The following terms and phrases are used in this chapter.

Arc-Reflecting: A test set application wherein a test signal is bounced or reflected from a fault or short circuit.

Characteristic Impedance: The AC resistance a cable exhibits to the passage of a pulse.

CRO: Abbreviation for cathode ray oscilloscope. A CRO is a vacuum device used to display electrical signals or other information. The screen of a television set is a CRO.

Fault: A short circuit or another flaw in a power system conductor.

OTDR: Optical time domain reflectometer. An OTDR is an instrument that applies a light pulse to an optical fiber. The OTDR then displays the waves that are reflected from the various faults and imperfections of the fiber. (See TDR for the electrical cable equivalent.)

TDR: Time domain reflectometer. A TDR is an instrument that applies a low-energy voltage pulse to a cable. The TDR then displays the waves that are reflected from various faults and imperfections in the cable.

Thumper: A test instrument that applies a high-energy voltage pulse to an electrical cable that has failed.

PURPOSE OF FAULT LOCATION

Power cables are subject to failure. A common practice has been to allow up to three splices in a cable before replacement. Before a cable can be repaired, the failed section must be accurately located. When the cable is aboveground, as in cable tray installation, locating the short circuit may be relatively easy. Buried cable or cable installed in conduit is a different matter.

A fault location technique needs to satisfy the following requirements:

- The fault must be located within a close enough tolerance that it can be found and repaired with a minimum amount of digging and confusion.
- The method used should be as nondestructive as possible. Minimum additional damage should be inflicted on the cable during the fault location process.

METHODS (ELECTRICAL CABLE)

Thumping

Thumping is the method of choice for locating high-resistance faults. The basic principle in thumping involves impressing a repetitive, high-energy pulse to the faulted line. The pulse is generated by rapidly discharging a capacitor to the cable. Because of the nature of the circuits involved, voltage peaks of as much as four times the thumper output voltage will be generated by the pulses that travel down the cable.

When a pulse reaches the fault, the excessive voltage will cause the faulted area to break down and arc. The fault may then be located in one of the following ways:

- If the cable is direct buried, the fault can often be located by simply walking along the cable route until the vibration of the arc is felt in the ground.

- Audio sensing equipment may be used if the arc is not of sufficient strength to be felt.

There are several advantages and disadvantages to thumping. The advantages include:

- Thumping is relatively simple and straightforward. Little specialized technical skill is needed to locate the fault.
- Thumping will locate even very high-resistance faults.

The disadvantages include:

- Because adjacent cables may be damaged by the thump, thumping may not be advisable for cable in cable trays or in multiple cable conduits.
- Some faults are very stubborn and may require hundreds or thousands of thumps to break the fault down.
- Because of the very high voltages and very rapid rise times, cable insulation is stressed to the maximum during the thumping process. If the thumping action takes too long, good insulation may be compromised.
- Acoustical echoes may make the location difficult to discern, even when the fault breaks down.

Despite the disadvantages, thumping is still the most popular of all the fault-location methods.

TDR (ELECTRICAL CABLE)

Power cable falls into the general category of a transmission line. At very high frequencies, all transmission lines possess an electrical property known as characteristic impedance. The characteristic impedance of a given cable is a function of the cable material and its dimensions. If the material or the dimensions change, the characteristic impedance changes. A splice, a fault, a bend, a transformer, or another discontinuity in a cable will have a different characteristic impedance than the cable itself.

Time domain reflectometry takes advantage of the fact that a voltage pulse will create electrical echoes every time it passes an area with a different characteristic impedance. In this method, a low-energy, high-frequency pulse is applied to the

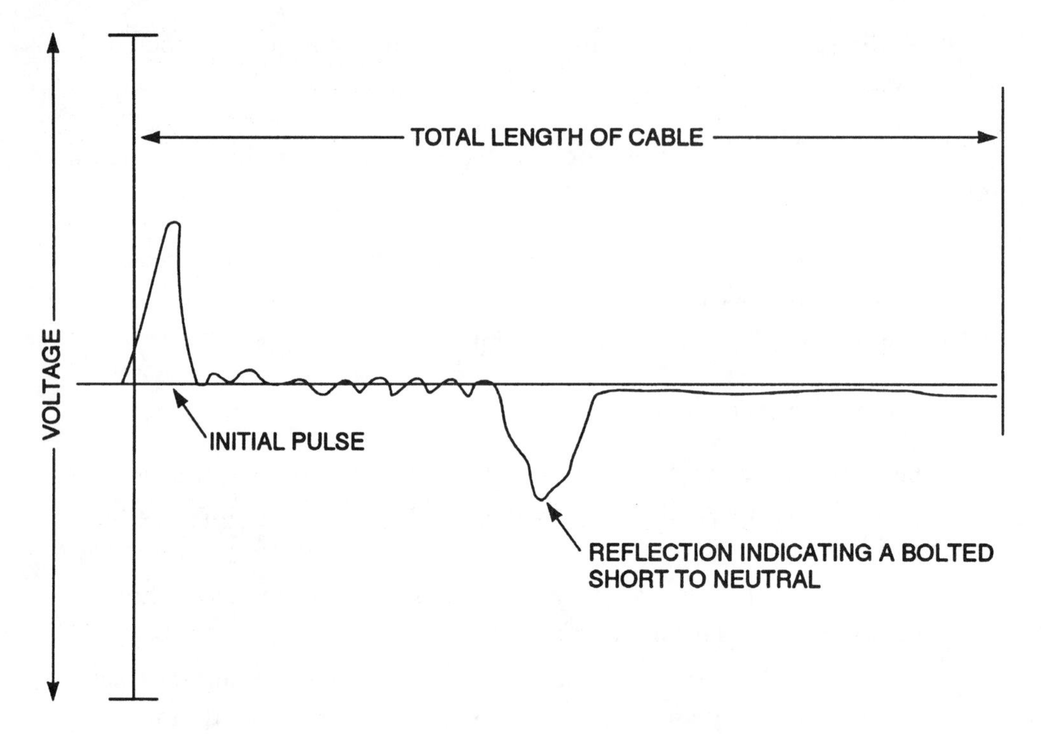

Figure 5–1 Typical layout for TDR display.
(Courtesy of Cadick Corporation)

cable. When the pulse reaches a point of irregularity, all or part of the energy is reflected back toward the input end of the cable.

At the input end of the cable, the time domain reflectometer uses a cathode ray oscilloscope (CRO) to display the voltages it detects. Figure 5–1 shows the general appearance of such a display.

The horizontal sweep control is adjusted so that the entire length of the cable is displayed on the CRO. The vertical sensitivity is adjusted so that the voltage pulses show clearly on the screen.

The initial pulse shows on the screen at the origin. The return pulse, if any, will reach the instrument at the same moment the horizontal sweep reaches the point on the screen that represents the point from which the return pulse originated.

Different types of irregularities will produce differently shaped return pulses. Figure 5–2 shows examples of six different traces caused by various types of problems. This partial list represents the most common problems encountered.

The advantages of TDR include:

- Is nondestructive.
- Generally provides a much more precise location of the fault.

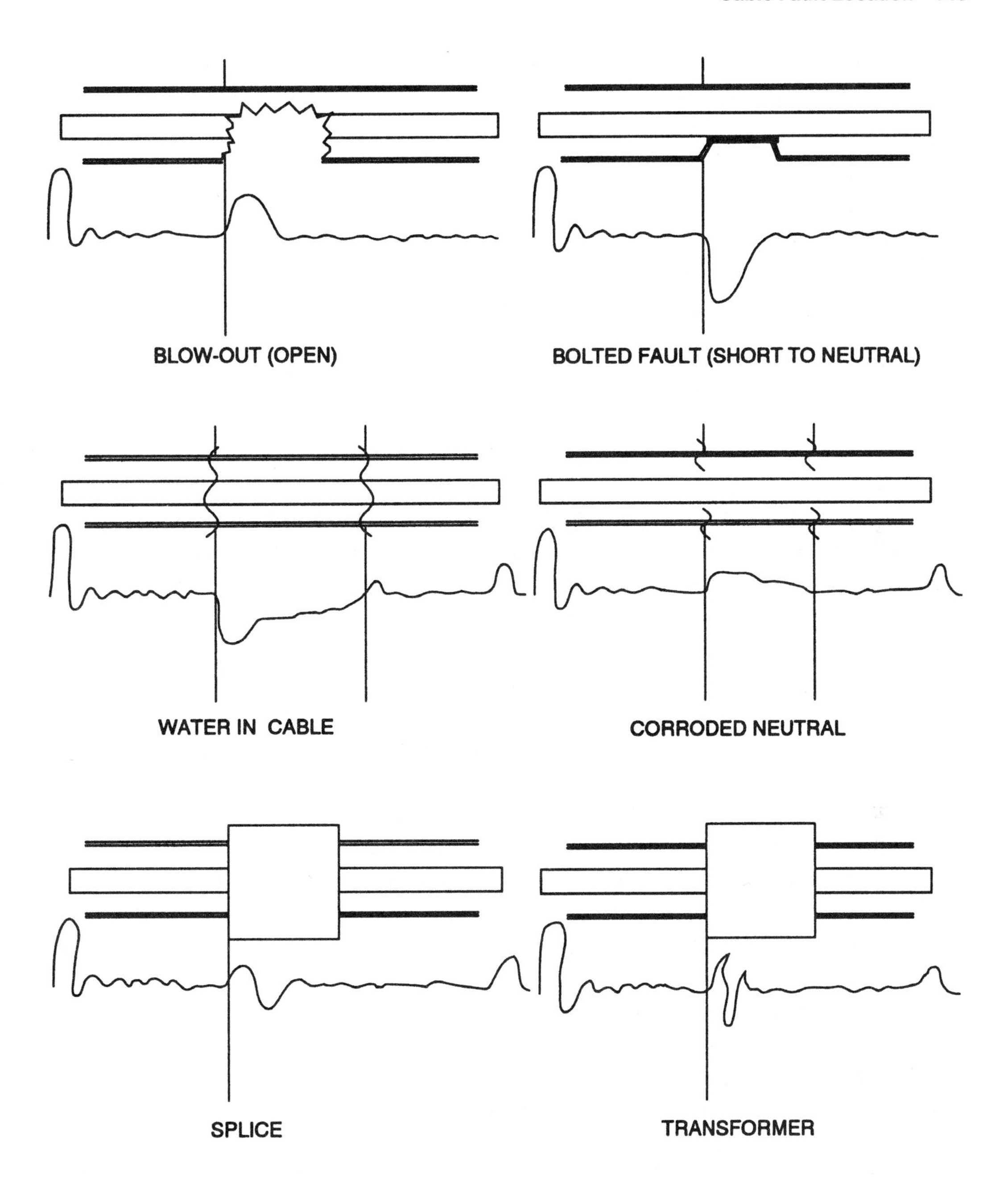

Figure 5–2 Typical signatures for power utility applications.
(Courtesy of AVO, International)

- May allow the user to determine the type and degree of short circuit, thereby allowing better splice planning.

The disadvantages of TDR include:

- Is extremely sensitive and picks up even minor problems. May make fault location difficult.
- Does not readily recognize high-resistance short circuits.
- The CRO may be difficult to read and generally requires an experienced operator.

Arc Reflection Methods

At least one manufacturer has introduced a test method that operates using a combination of thumping and time domain reflectometry. This system will operate as a TDR for all the types of faults the TDR can readily locate. For high-resistance short circuits, the operator can select a mode called arc reflection. In this mode, the test equipment applies a voltage pulse to the cable in much the same way a thumper does, but the arc reflection system uses only the minimum amount of energy required to break the fault down into an arc.

While the arc-creating voltage pulse is applied, the test equipment applies a high-frequency, low-energy pulse. The low-energy pulse reflects off the arc and is shown on the CRO in much the same way were a low-resistance fault present.

The major disadvantage of the TDR system is its inability to locate high-resistance faults. The arc-reflection method eliminates this problem by using a voltage pulse to break down the fault. The major disadvantage to thumping is the likelihood that the cable may be damaged by the high-energy pulses. The arc-reflection method minimizes this problem by using only the minimum amount of energy required to create the arc.

METHODS (OPTICAL FIBER)

OTDR

The OTDR is an optical version of the TDR discussed previously in this chapter. Most OTDRs are extremely automated devices that can perform some or all of the following functions:

- Locate faults, splices, and other fiber discontinuities.
- Measure attenuation of the fiber.

- Compare two or, in some cases, more fibers.

While the optical power meter discussed in Chapter 4 is a good and useful tool, the OTDR is the choice of most professionals involved in installing and troubleshooting fiber optic cable.

TEST EQUIPMENT

Figures 5–3 and 5–4 show examples of typical TDRs. The hand-held model is a convenient portable package that is used for lighter duty applications. The larger model is a more deluxe model with built-in memory and a full range of calibration and adjustment features.

Figure 5–5 depicts a thumper. This unit applies a high-energy pulse to the cable and causes it to break down.

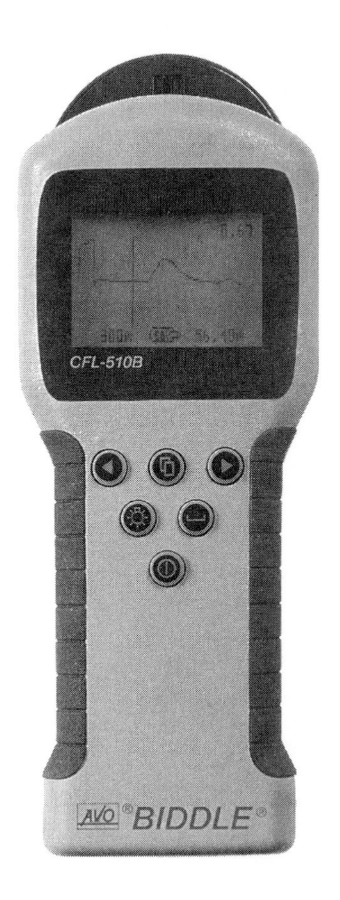

Figure 5–3 Hand-held TDR.
(Courtesy of AVO, International)

Figure 5–6 shows an arc reflection unit. It may be considered to be a thumper in the same package with a TDR. The energy pulse that is produced is high enough to cause a low resistance arc to form, yet low enough to minimize or avoid any additional damage to the cable. Simultaneously, a low-energy, high-frequency pulse is placed on the cable. This pulse reflects off the arc and puts a "blip" on the screen.

Figure 5–7 is an optical time domain reflectometer. This particular unit is very portable, yet has the ability to display and perform the variety of tests described earlier.

Testing Patch Cables (FOTP-171)

This procedure includes four methods, with three optional procedures for each method. We will restrict our discussion to Method B, which is the best suited to premises cabling and is the method typically used by cable assemblers.

This procedure (Figure 5–6) uses the substitution method to evaluate attenuation in a length of fiber. Power through two jumper cables is measured to zero on the meter. The cable assembly to be tested is then inserted between the two jumper fibers. The new reading is the fiber attenuation. Notice that the test requires a mode

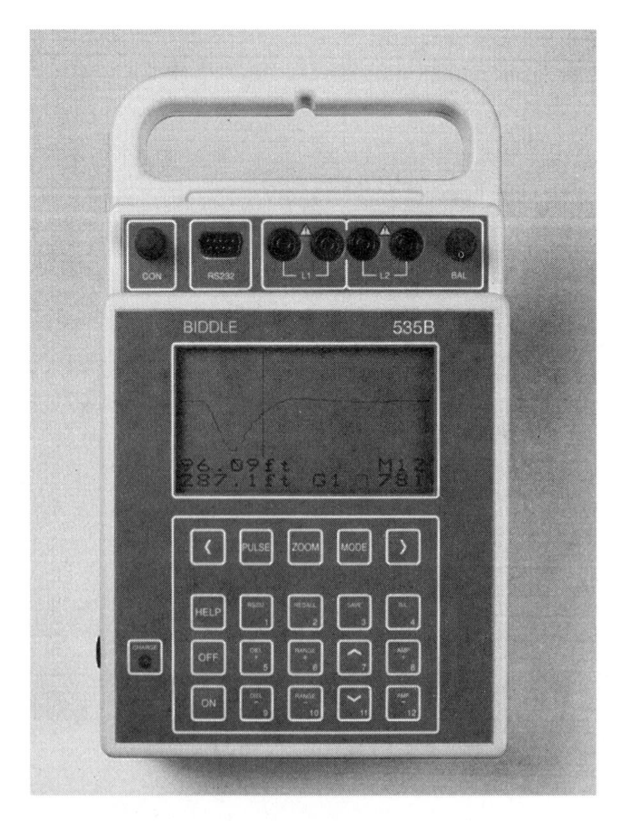

Figure 5–4 Typical TDR unit.
(Courtesy of AVO, International)

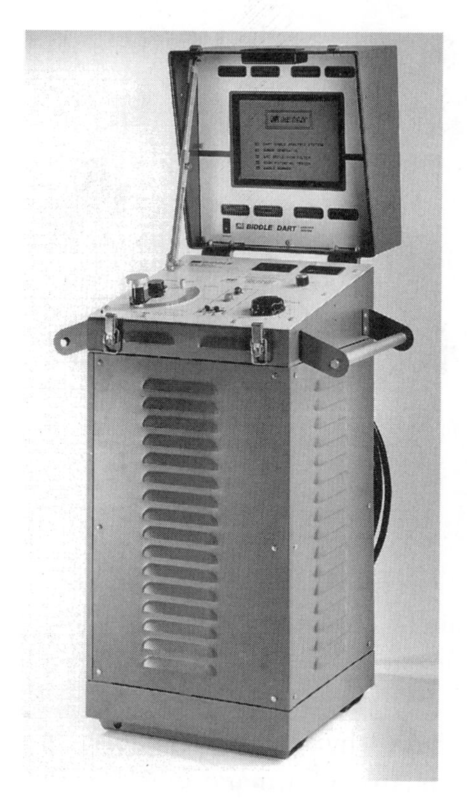

Figure 5–5 Typical thumper unit.
(Courtesy of AVO, International)

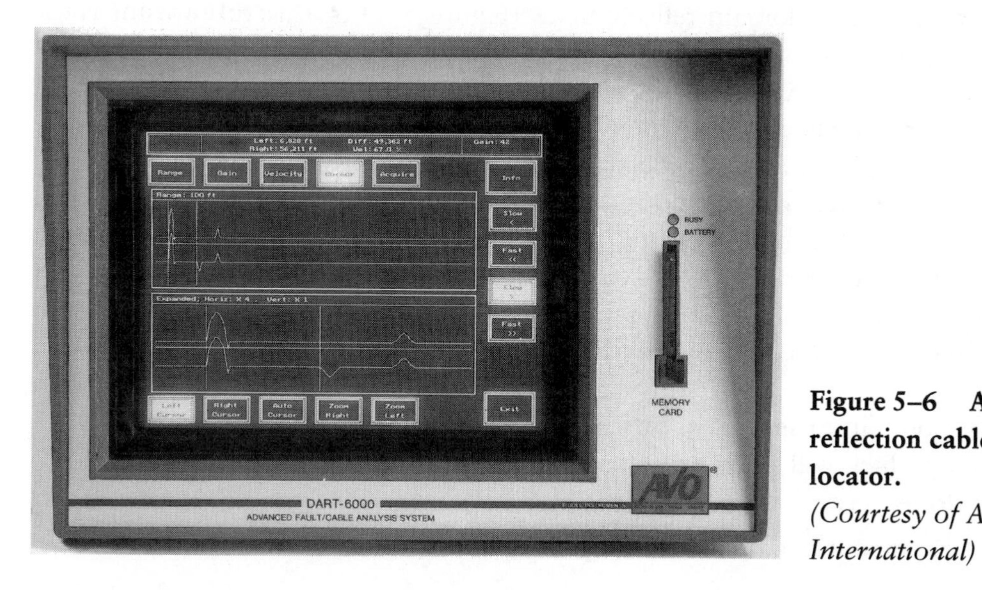

Figure 5–6 Arc reflection cable fault locator.
(Courtesy of AVO, International)

Figure 5–7 Optical time domain reflectometer.
(Courtesy of Photon Kenetics)

filter on the launch end to simulate EMD in the short length of jumper cable. This can be achieved by wrapping the cable five turns around a mandrel to mix the modes.

Optical Time-Domain Reflectometry

An optical time-domain reflectometer (Figure 5–7) is a useful tool for troubleshooting. We discussed the uses of time-domain reflectometry in testing UTP links. Certification tools use TDRs to measure cable lengths. OTDR—optical TDR—is more widely used than metallic TDR. The principles are the same: a pulse is sent down the fiber. Energy reflecting from imperfections and changes in refractive index of the path are detected and displayed. Changes in the index of refraction of a fiber are analogous to changes in impedance in a copper cable: both cause reflections that can be used to get a clear picture of the link.

Most TDRs are highly automated and sophisticated, allowing you to zoom into specific areas for a close inspection. Among capabilities are these:

- Zoom in to specific events like connectors, splices, and faults (all of which will show pronounced bumps in the trace) for a closeup look.
- Zoom out to gain an overall picture of the link.

PART TWO

Chapter 6

Cable Application Guide

Cable Types

Armored Cable (Type AC)
Fluorinated Ethylene Propylene (Types FEP and FEPB)
Flat Conductor Cable (Type FCC)
Integrated Gas Spacer Cable (Type IGS)
Metal-Clad Cable (Type MC)
Mineral Insulated Cable (Type MI)
Machine Tool Wire (Type MTW)
Medium-Voltage Cable (Type MV)
Nonmetallic-Sheathed Cable (Types NM, NMC, and NMS)
Perfluoroalkoxy (Types PFA and PFAH)
Heat-Resistant Rubber (Types RH, RHH, RHW, and RHW-2)
Silicone Asbestos (Type SA)
Service Entrance Cable (Types SE and USE)
Synthetic Heat Resistance (Type SIS)
Thermoplastic Asbestos (Type TA)
Thermoplastic and Fibrous Outer Braid (Type TBS)
Tray Cable (TC)
Extended Polytetrafluoroethylene (Type TFE)
Thermoplastic (Types THHN, THHW, THW, THWN, and TW)
Underground Feeder and Branch Circuit Cable (Type UF)
Varnished Cambric (Type V)
Cross-Linked Polymer (Types XHHW and XHHW-2)
Modified Ethylene Tetrafluoroethylene (Types Z and ZW)
Remote Control, Signaling, and Power-Limited Circuit Cable (CL*x*)
Power Limited Tray Cable (PLTC)
Nonpower Limited and Power-Limited Fire Protective Signaling Circuit Cable (NFPL and FPL)

Optical Fiber Cable (OFC and OFN)
Communications and Multipurpose Cables (CM*x* and MP*x*)

INTRODUCTION

One of the most often encountered problems in the field is when, where, and how to use the different types of cables and cable insulations available. This chapter is designed as an application guide and includes information concerning the proper way to use these various cable and insulation types. This chapter includes information about UL-listed cable types and their insulation types.

UL-listed cables are assemblies that are manufactured for specific applications and include types AC, FCC, IGS, MC, MI, MTW, MV, NM, NMC, NMS, SE, USE, TC, and UF.

Insulation materials used for the various UL-listed cable types are found in Table 310–13 of the *NEC*® and include types FEP, FEPB, PF, PFAH, RH, RHH, RHW, RHW-2, SA, SIS, TA, TBS, TFE, THHN, THHW, THW, THWN, TW, V, XHHW, XHHW-2, Z, and ZW.

In this second edition of *Cables and Wiring,* we have added communications and signaling cables to Chapter 6. The new NEC cable types include CL1, CL2, CL3, PLTC, FPL, OFC, OFN, CM, and MP. Each type is included with its plenum, riser, commercial, and residential UL equivalents. Because power cable and communications cable have somewhat different requirements, a new section has been added for communications cable.

Because the *NEC*® requires that some of the specified cable types be made of certain types of insulation, principally fixture wire, this edition also includes a copy of *NEC*® Table 402–3, which identifies each fixture wire type. This table can be found at the end of this chapter in Appendix A.

NOTE:

All of the cable types listed in this chapter are available from manufacturers. Some, such as varnished cambric (V), are not specified for new installations and may be available only from specialty manufacturers. Cable types that previously used asbestos, such as FEBP and SA, are being replaced with other materials, such as glass fiber.

The splicing and termination of these cable types is covered in Part One of this handbook. While special requirements have been noted in many cases, the user must refer to the manufacturer's recommendations and application *NEC*® sections for detailed information and regulations.

ARMORED CABLE (TYPE AC)

Description

General. Type AC cable is a fabricated assembly of insulated conductors in a flexible metal enclosure. Figure 6–1 shows a type of armored cable.

Types. Armored cable is available in the following types:

AC: No suffix indicates 60° Celsius insulation. Conductors have thermoplastic insulation.
ACH: Suffix indicates 75° Celsius insulation.
ACHH: Suffix indicates 90° Celsius insulation.
ACL: Suffix indicates lead covering.
ACT: Suffix indicates thermosetting insulation.

Applications

Armored cable is used in applications in which the cable might be exposed to damaging conditions that cannot be minimized or eliminated by other types of systems.

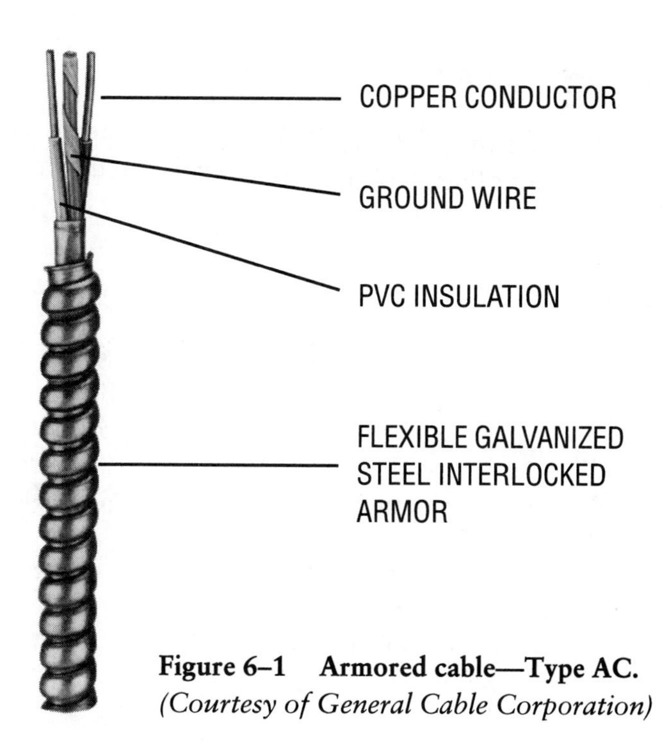

Figure 6–1 Armored cable—Type AC.
(Courtesy of General Cable Corporation)

Permitted	Not Permitted
Dry locations (Type AC)	Wet locations (Type AC)
Under plaster extensions	Theaters (518)
Embedded in plaster finish	Motion picture studios
On brick or other masonry	Hazardous locations (501 and 504)
	Corrosive fumes or vapors
Wet locations (Type ACL only)	Cranes or hoists (610)
	Storage battery rooms
	Hoistways or on elevators (620)
	Commercial garages (511)

NOTE:

Numbers in parentheses indicate articles in the *NEC®* where additional or explanatory information may be found.

Sizes

Copper	Aluminum or Copper-Coated Aluminum
#14 AWG up to #1 AWG	#12 AWG up to #1 AWG

Temperature Ranges

Armored cable is available in temperature ranges **up to 90° Celsius.** The allowed application temperature is determined by the type of insulation used on the conductors.

Voltage Ranges

Armored cable is intended for use in **less than 2,000 volt** applications.

Ampacities

Armored cable ampacities are determined using standard methods from the *NEC®*.

Handling and Care

Receiving and Handling.

- Examine the protective covering for evidence of damage.
- Do not allow equipment to touch or damage cable surface or protective wrap.
- If a forklift is used, the forks must be long enough to contact both reel flanges.
- If a crane is used, a shaft through the arbor hold or flange cradles must be employed for the lift.
- If an inclined ramp is used for unloading, it must be wide enough to contact both flanges completely. The reel should be stopped at the bottom by using the flanges, not the cable surface.
- Never drop reels on the ground.

Storage.

- Unjacketed armored cable should be stored indoors to avoid corrosion.
- Reels should be stored on a hard, clean surface, not bare earth.
- Store away from open fires or heat sources.
- Cable ends should be resealed as soon as a cable section has been removed.

Receiving Tests. Cable should be tested according to the insulation type and the methods outlined in Chapter 4 of this handbook.

Installation

Securing and Supporting. Armored cable shall be secured and supported as follows:

- Approved staples, straps, hangers, or similar fittings, as shown in Figure 6–2.
- When mounted on studs, joists, and rafters, Type AC cable shall be installed as shown in Figure 6–3.
- When passed through bored holes, the edge of the hole shall be no less than 1¼ inches from the edge of the stud, joist, or rafter.
- In the event that 1¼ inches cannot be maintained, as mentioned previously and shown in Figure 6–3, a steel spacer or sleeve must be used to protect the cable.

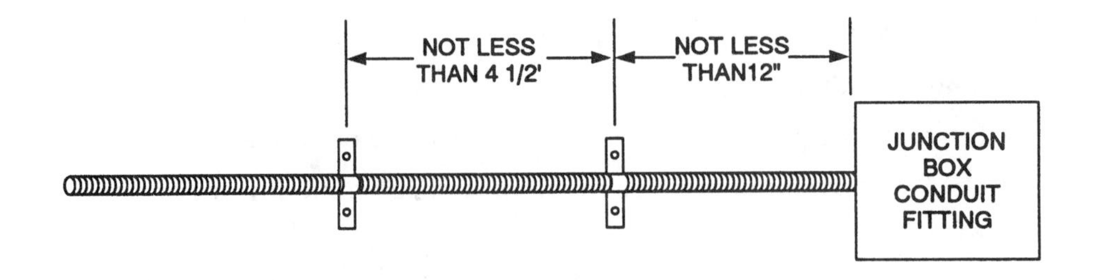

Figure 6–2 Securing armored cable.
(Courtesy of Cadick Corporation)

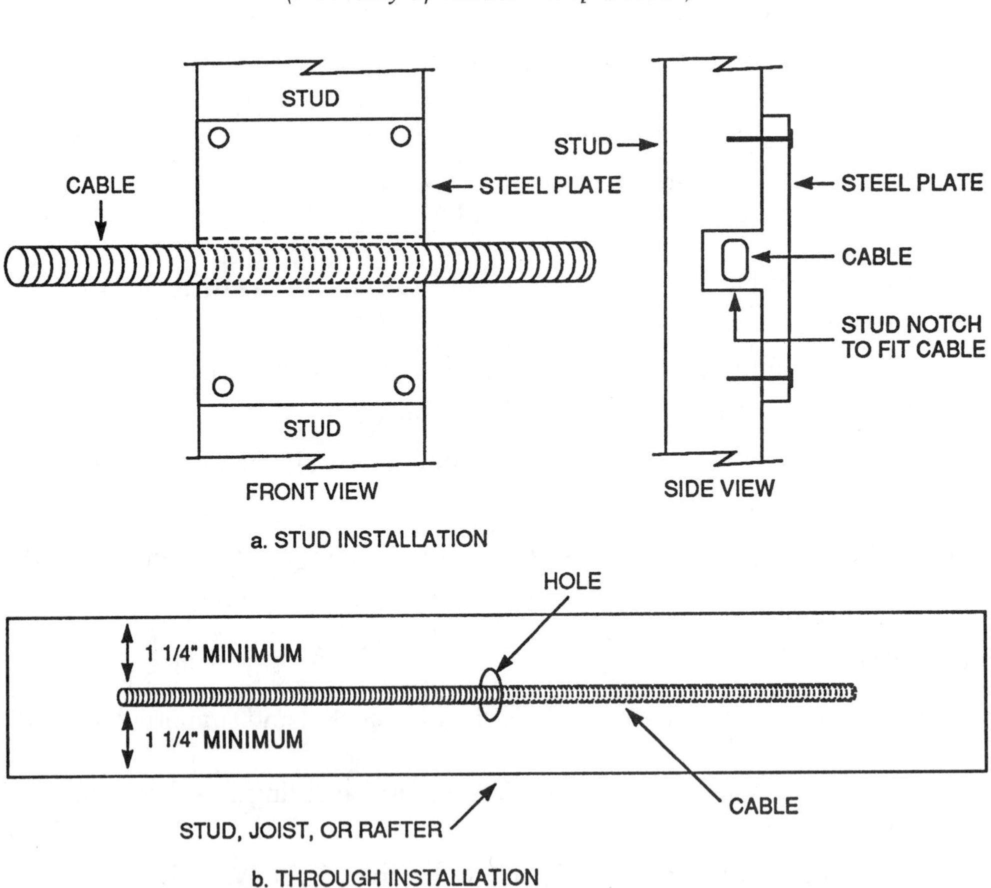

Figure 6–3 Installing cable on studs, joists, and rafters.
(Courtesy of Cadick Corporation)

- Armored cable installed in accessible attics shall be secured, as shown in Figure 6–4.

Termination and Splicing. When armored cable is terminated, a fitting must be used to protect the insulation from abrasion and an insulating bushing must be provided between the conductors and the armor.

If the armor is removed for the purpose of a splice or termination, the splice or termination must be made in a box or fitting approved for that purpose.

Bending Radius.

- Not less than five times the cable diameter.
- In no case shall the bending radius be so small as to damage the cable.

FLUORINATED ETHYLENE PROPYLENE (TYPES FEP AND FEPB)

Description

General. Types FEP and FEPB are insulation types that are used in high-temperature applications.

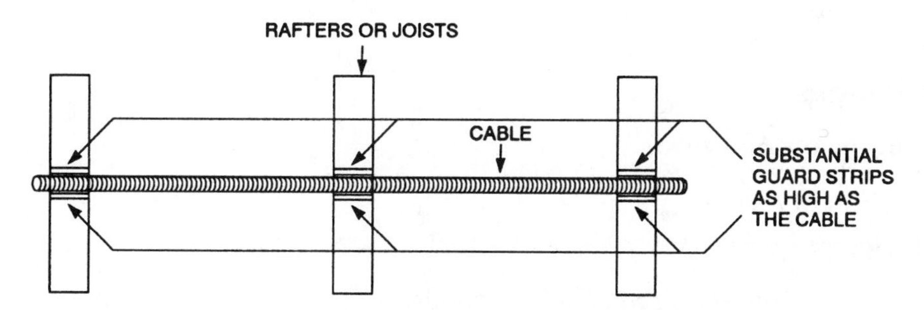

a. OVER JOISTS OR RAFTERS IN ATTICS ACCESSIBLE BY LADDER OR STAIRS, AS WITHIN 6 FEET OF SCUTTLE HOLE OR ATTIC ENTRANCE

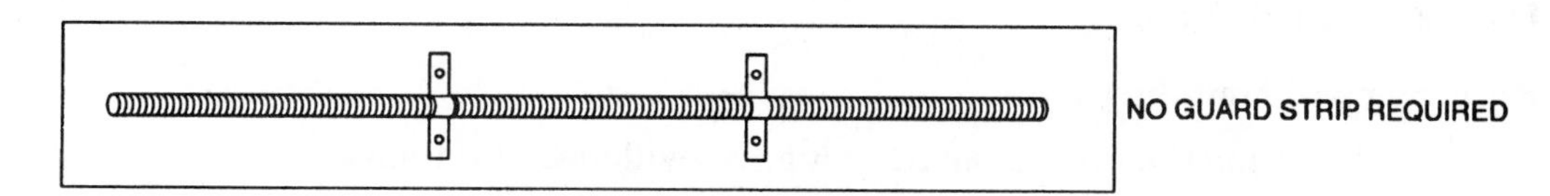

b. ALONG THE SIDE OF JOISTS, STUDS, OR RAFTER

Figure 6–4 Installing armored cable in accessible attics.
(Courtesy of Cadick Corporation)

Types. Fluorinated ethylene propylene cable is available in the following types:

FEP Dry and damp locations up to 90° Celsius.
FEP Dry locations up to 200° Celsius; special applications.

Applications

FEP	**FEPB**
Dry and damp locations up to 90° Celsius	Dry locations up to 200° Celsius; special applications only

Sizes

FEP From **#14 AWG up to #2 AWG.**
FEPB From **#14 AWG up to #8 AWG** with glass braid.
From **#6 AWG up to #2 AWG** with asbestos or other suitable material.

Temperature Ranges

Up to 200° Celsius; see applications table.

Voltage Ranges

Circuits **below 600 volts.**

Ampacities

Determined using standard *NEC*® methods.

Handling and Care

Receiving and Handling.

- Examine the protective covering for evidence of damage.
- Do not allow equipment to touch or damage cable surface or protective wrap.
- If a forklift is used, the forks must be long enough to contact both reel flanges.

- If a crane is used, a shaft through the arbor hold or flange cradles must be employed for the lift.
- If an inclined ramp is used for unloading, it must be wide enough to contact both flanges completely. The reel should be stopped at the bottom by using the flanges, not the cable surface.
- Never drop reels on the ground.

Storage.

- Reels should be stored on a hard, clean surface, not bare earth.
- Store away from open fires or heat sources.
- Cable ends should be resealed as soon as a cable section has been removed.

Receiving Tests. Cable should be tested according to the insulation type and the methods outlined in Chapter 4 of this handbook.

Installation

Securing and Supporting. Cable shall be secured and supported as follows:

- Approved staples, straps, hangers, or similar fittings, as shown in Figure 6–21.
- When mounted on studs, joists, and rafters, cable shall be installed as shown in Figure 6–22.
- When passed through bored holes, the edge of the hole shall be no less than 1¼ inches from the edge of the stud, joist, or rafter.
- In the event that 1¼ inches cannot be maintained, as mentioned previously and shown in Figure 6–3, a steel spacer or sleeve must be used to protect the cable.
- Cable installed in accessible attics shall be secured, as shown in Figure 6–23.

Termination and Splicing. Figure 6–24 shows the correct methods for termination of fluorinated ethylene propylene cable.

FLAT CONDUCTOR CABLE (TYPE FCC)

Description

General. Type FCC cable consists of three or more flat copper conductors placed edge-to-edge and separated and enclosed within an insulating assembly. Figure 6–5 shows a cross section of a typical flat conductor cable.

An FCC system is intended for use in branch circuits and is to be installed below carpet squares. It is installed as a complete system, including terminal boxes, receptacles, and wire (see Figure 6–6).

Types. N/A.

Applications

FCC cable is intended for use in branch circuits to provide a complete and accessible power system. FCC cable allows for easy upgrading of existing office branch circuit power systems.

Permitted	Not Permitted
Branch circuits	Outdoors or in wet locations
General purpose	Where subject to corrosive vapors
Appliance	Hazardous (classified) locations
Individual	Residential buildings
Floors	School buildings
Concrete	Hospital buildings
Ceramic	
Composition	
Wood	
Walls	
In surface metal raceways	
Damp locations	
Heated floors	
Must be made from materials approved for floors above 30° Celsius	

Sizes

Three, four, and five conductors shall be arranged in each system. Wire sizes shall be of sufficient ampacity, as described later.

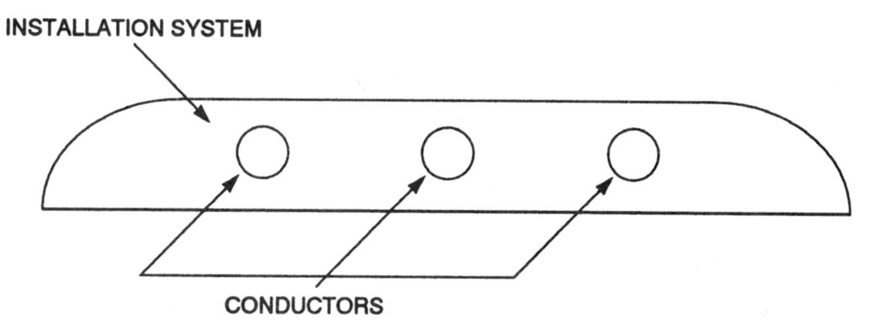

Figure 6–5 Cross section of typical flat conductor cable.
(Courtesy of Cadick Corporation)

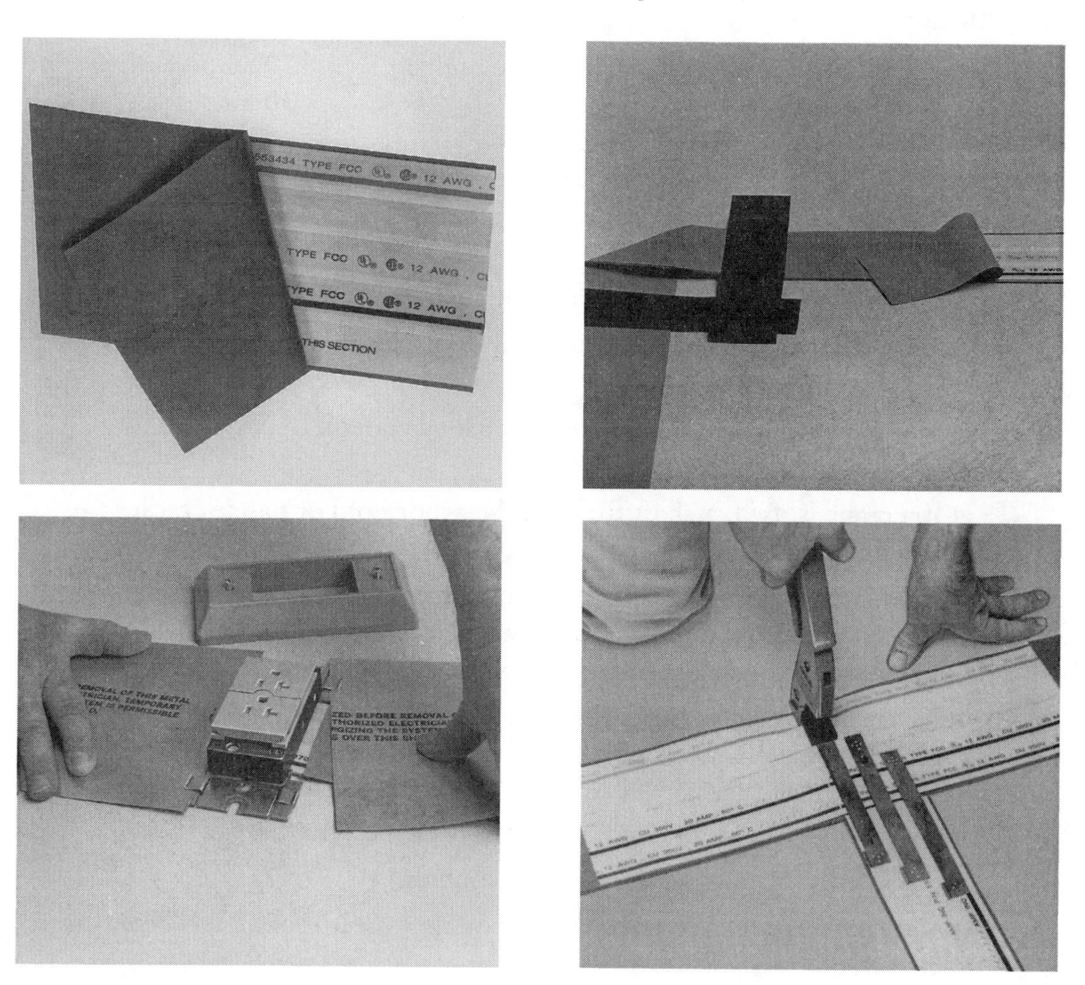

Figure 6–6 Elements of a flat conductor cable system.
(Courtesy of AMP, Incorporated, Harrisburg, Pennsylvania)

Temperature Ranges

FCC systems shall be installed at operating temperatures that are allowed according to the type of insulation, generally **60, 75, or 90° Celsius.**

Voltage Ranges

FCC cable shall be used in systems with **300 volts maximum** between ungrounded conductors and **150 volts maximum** between ungrounded and grounded conductors.

Ampacities

General-purpose and appliance branch circuits:	**20 amperes**
Individual branch circuits:	**30 amperes**

Handling and Care

Receiving and Handling.

- Examine the protective covering for evidence or damage.
- Do not allow equipment to touch or damage cable surface or protective wrap.
- If a forklift is used, the forks must be long enough to contact both reel flanges.
- If a crane is used, a shaft through the arbor hold or flange cradles must be employed for the lift.
- If an inclined ramp is used for unloading, it must be wide enough to contact both flanges completely. The reel should be stopped at the bottom by using the flanges, not the cable surface.
- Never drop reels on the ground.

Storage.

- Reels should be stored on a hard, clean surface, not bare earth.
- Store away from open fires or heat sources.
- Cable ends should be resealed as soon as a cable section has been removed.

Receiving Tests. Cable should be tested according to the insulation type and the methods outlined in Chapter 4 of this handbook.

Installation

Securing and Supporting. FCC systems cable shall be secured and supported as follows and as shown in Figure 6–7.

- A metallic or nonmetallic bottom shield is mounted to the floor using an approved adhesive.
- The FCC cable and other components are placed on top of the shield.
- A metallic top shield is placed over the FCC cable.
- The carpet square (36 square inch maximum) is placed over the top shield.
- If any portion of the FCC system is more than 0.09 inch above the floor, it must be tapered or feathered.
- Crossing of more than two FCC cables at one point is not permitted.
- When any crossing occurs, the cables shall be separated by a grounded metal shield.

Termination and Splicing. All terminations and splices shall be made with equipment and materials specifically designed for FCC systems. Connections shall be installed such that electrical continuity, insulation, and sealing against dampness and liquid spillage are provided. Refer to the manufacturer's instructions for specific termination and splicing information.

Bending Radius. N/A.

INTEGRATED GAS SPACER CABLE (TYPE IGS)

Description

General. Figure 6–8 shows the general construction of Type IGS cable. The following descriptions explain the basic construction.

Aluminum rods: 250 kcmil (aluminum) conductors.

Strand shield: Aluminum tape shield used to reduce voltage gradient stress and provide support for the conductors.

Paper spacer: Provides support and insulation between the strand shield and the shield tape.

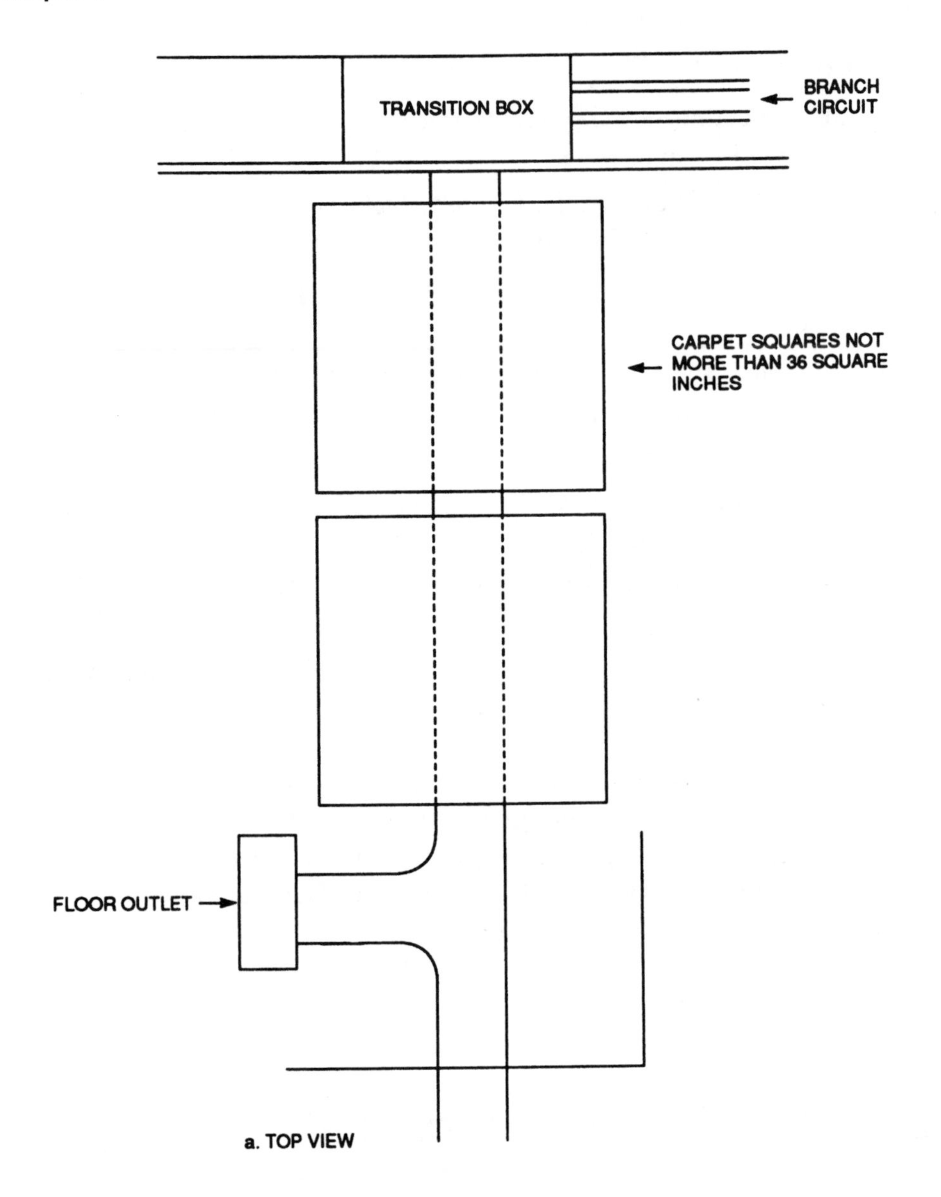

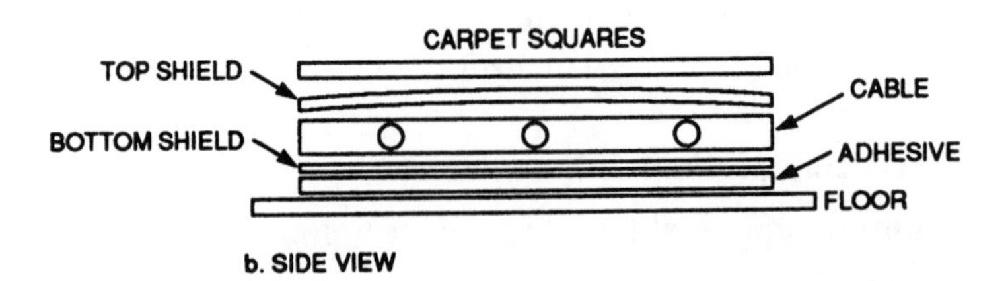

Figure 6–7 Installation of a flat conductor cable system.
(Courtesy of Cadick Corporation)

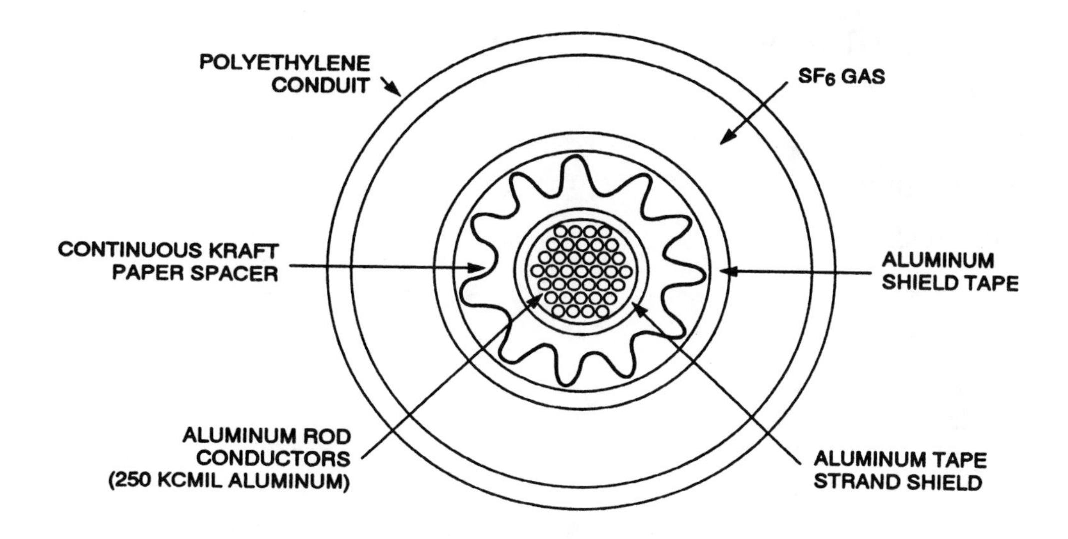

Figure 6–8 Cross section of Type IGS cable.
(Courtesy of Cadick Corporation)

Shield tape: Provides voltage gradient stress reduction and neutral paths.

SF_6 gas: Insulation.

Conduit: Mechanical protection and containment for the SF_6 gas.

Types. N/A.

Applications

Type IGS cable is used for underground feeder applications.

Permitted	**Not Permitted**
Underground feeders	Interior wiring
Direct burial	Exposed in contact with buildings
Service entrance (USE)	
Feeders	
Branch circuits	

Sizes

250 kcmil up to 4,750 kcmil.

Temperature Ranges

N/A.

Voltage Ranges

0 to 600 volts.

Ampacities

kcmil	amperes
250	119
500	168
750	206
1,000	238
1,250	266
1,500	292
1,750	315
2,000	336
2,250	357
2,500	376
3,000	412
3,250	429
3,500	445
3,750	461
4,000	476
4,250	491
4,500	505
4,750	519

Handling and Care

Receiving and Handling.

- Examine the protective covering for evidence of damage.
- Do not allow equipment to touch or damage cable surface or protective wrap.
- If a forklift is used, the forks must be long enough to contact both reel flanges.
- If a crane is used, a shaft through the arbor hold or flange cradles must be employed for the lift.

- If an inclined ramp is used for unloading, it must be wide enough to contact both flanges completely. The reel should be stopped at the bottom by using the flanges, not the cable surface.
- Never drop reels on the ground.

Storage.

- Reels should be stored on a hard, clean surface, not bare earth.
- Store away from open fires or heat sources.
- Cable ends should be resealed as soon as a cable section has been removed.

Receiving Tests. Cable should be tested according to the insulation type and the methods outlined in Chapter 4 of this handbook.

Installation

Securing and Supporting. IGS cable is intended for direct burial installation.

Termination and Splicing. Terminations and splices are required to be made with approved fittings suitable for maintaining the pressure of the SF_6 gas. A valve and cap is required for each length of cable. The valve and cap is used to measure gas pressure and to inject gas when required.

Bending Radius. IGS cable is allowed to have no more than four quarter bends (360° total). Table 6–1 summarizes the minimum bending radius requirements.

Table 6–1 Minimum bending radii for IGS cable.

Conduit Trade Size (Inches)	Minimum Bending Radius (Inches [mm])
2	24 (610)
3	35 (889)
4	45 (1,140)

METAL-CLAD CABLE (TYPE MC)

Description

General. Type MC cable is a fabricated assembly of one or more individually insulated conductors in a metallic sheath. Figure 6–9 shows a type of metal-clad cable.

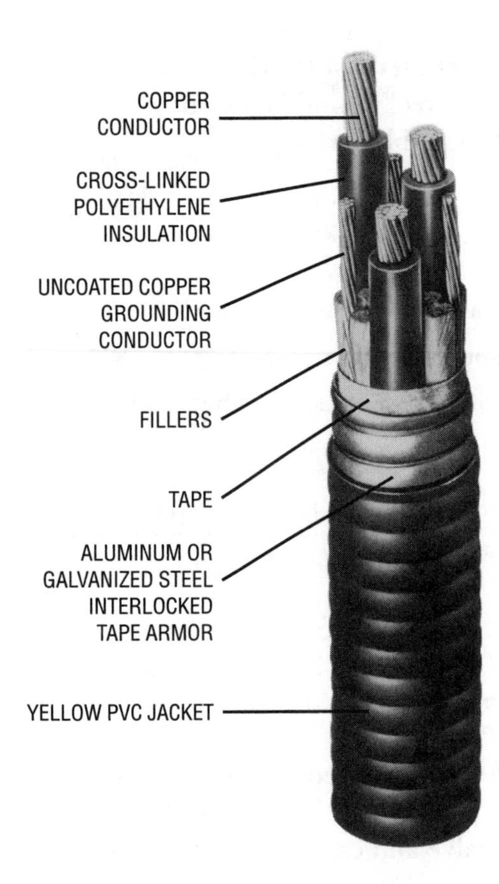

Figure 6–9 Metal-clad cable.
(Courtesy of General Cable Corporation)

Types. Metal-clad cable is designated as type MC, with the following additions:

CS Copper sheath.
ALS Aluminum sheath.

Applications

Permitted

Services, feeders, and branch circuits
Power, lighting, control, and signal
Indoors or outdoors
Exposed or concealed
Direct burial when so identified
Cable tray
Approved raceway
Open runs of cable

Not Permitted

Destructive corrosive conditions, except when the specific cable is designed for such an application

Permitted	**Not Permitted**
Aerial cable with messenger	
Hazardous locations (501, 502, 503, 504)	
Dry locations	
Wet locations under any of the following conditions:	
Metallic covering impervious to moisture	
Lead sheath or other jacket	
Insulated conductors suitable for wet applications	

NOTE:

Numbers in parentheses indicate articles in the *NEC*® where additional or explanatory information may be found.

Sizes

Copper	**Aluminum or Copper-Coated Aluminum**
#18 AWG and larger	**#12 AWG and larger**

Temperature Ranges

Metal-clad cable is available in temperature ranges **up to 90° Celsius.** The allowed application temperature is determined by the type of insulation used on the conductors.

Insulations with higher allowable temperatures may be used in special applications.

Voltage Ranges

Metal-clad cable is intended for use in circuits **up to 15,000 volts.**

Ampacities

Metal-clad cable ampacities are determined using standard methods from the *NEC*®.

Handling and Care

Receiving and Handling.

- Examine the protective covering for evidence of damage.
- Do not allow equipment to touch or damage cable surface or protective wrap.
- If a forklift is used, the forks must be long enough to contact both reel flanges.
- If a crane is used, a shaft through the arbor hold or flange cradles must be employed for the lift.
- If an inclined ramp is used for unloading, it must be wide enough to contact both flanges completely. The reel should be stopped at the bottom by using the flanges, not the cable surface.
- Never drop reels on the ground.

Storage.

- Unjacketed, metal-clad cable should be stored indoors to avoid corrosion.
- Reels should be stored on a hard, clean surface, not bare earth.
- Store away from open fires or heat sources.
- Cable ends should be resealed as soon as a cable section has been removed.

Receiving Tests. Cable should be tested according to the insulation type and the methods outlined in Chapter 4 of this handbook.

Installation

Securing and Supporting. Metal-clad cable shall be secured and supported as follows:

- Approved staples, straps, hangers, or similar fittings, as shown in Figure 6–10.
- In cable trays where approved for that application.
- Metal-clad cable may be direct buried, as per Figure 6–11.
- Metal-clad cable may be used for overhead or underground service entrance cable.

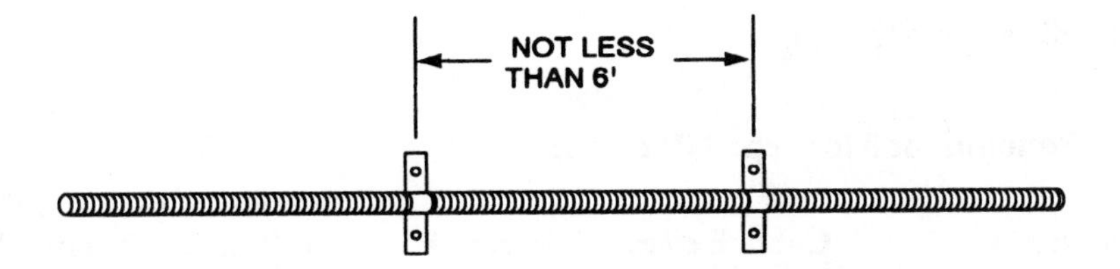

Figure 6–10 Securing metal-clad cable.
(Courtesy of Cadick Corporation)

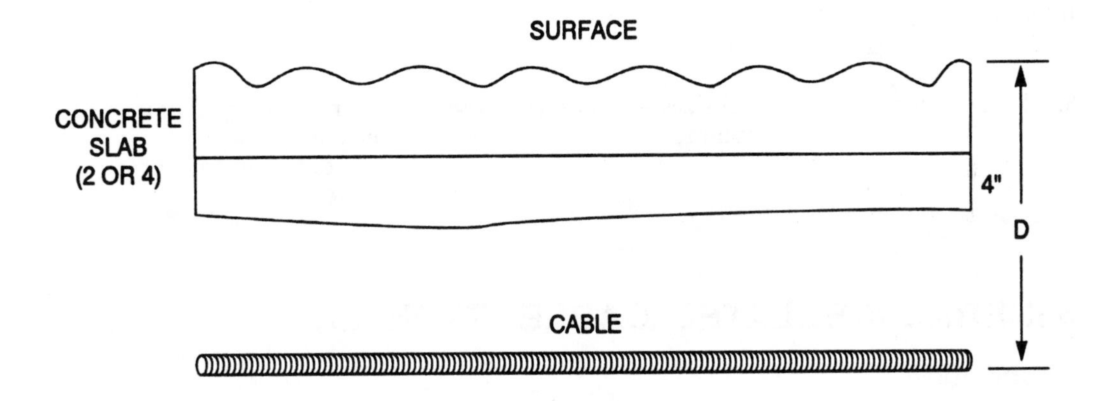

LOCATION	D (in Inches)
1. ALL LOCATIONS	24
2. BELOW 2" CONCRETE PAD IN A TRENCH	18
3. BENEATH A BUILDING (RACEWAY ONLY)	0
4. UNDER 4" CONCRETE SLAB (NO TRAFFIC, SLAB OVER HANG 6" MINIMUM)	18
5. STREET, HIGHWAYS, ETC.	24
6. ONE- AND TWO-FAMILY DWELLING DRIVEWAYS	18
7. AIRPORT RUNWAYS	18

Figure 6–11 Metal-clad direct burial cable.
(Courtesy of Cadick Corporation)

Termination and Splicing. Termination of Type MC cables should follow the procedures presented in Chapter 3. The metal sheath must be continuous and grounded. If a termination or splice is made at the junction with the cable system and a raceway system, it must be made in a box approved for that purpose.

Bending Radius. (see Table 6–2.)

Table 6–2 Bending radii for Type MC cable.

Cable Structure	Cable External Diameter	Minimum Bending Radius in Multiple of External Diameter
Smooth sheath	Less than 3/4 inch (19 mm)	10
	More than 3/4 inch (19 mm) but not more than 1 1/2 inch (38 mm)	12
	More than 1 1/2 inch (38 mm)	15
Interlocked-type armor or corrugated sheath	All sizes	7
Shielded conductors	All sizes—choose whichever is largest	12 times overall diameter of one conductor 7 times the overall cable diameter

MINERAL INSULATED CABLE (TYPE MI)

Description

General. Type MI cable consists of a copper conductor or conductors insulated with a highly compressed mineral insulation, usually magnesium oxide. The entire assembly is covered by a gastight and liquidtight copper or stainless steel sheath. Figure 6–12 shows the general construction of Type MI cable.

Types. N/A.

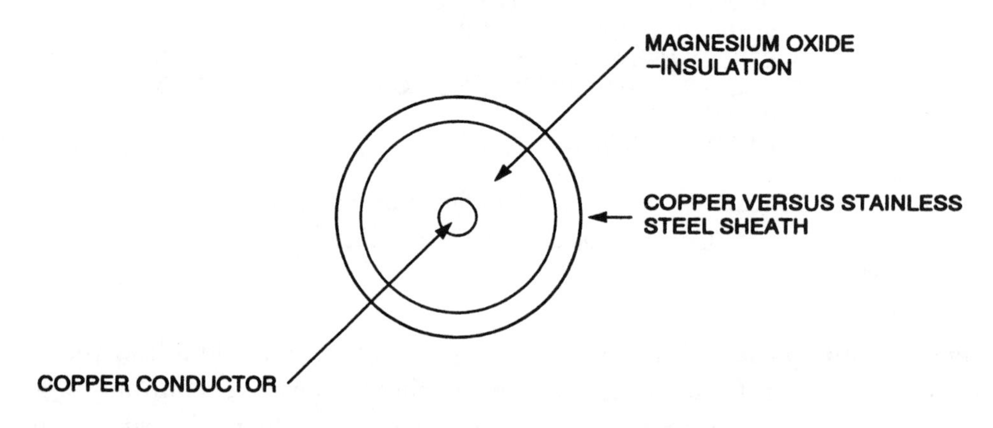

Figure 6–12 Mineral insulated cable.
(Courtesy of Cadick Corporation)

Applications

Type MI cable is used in applications in which high resistance to fire damage is necessary. The metallic sheath and the magnesium oxide insulation are capable of withstanding fire temperatures in excess of 1 hour.

Permitted	Not Permitted
Services, feeders, and branch circuits	Where exposed to damaging corrosion
Power, lighting, control, signal circuits	
Dry, wet, or continuously moist	
Indoors or outdoors	
Exposed or concealed	
Embedded in plaster, concrete, or other	
Above or below grade	
Any hazardous location	
Exposed to oil or gasoline	
Corrosion not deteriorating to sheath	
Underground runs	

Sizes

UL listings provide for the following cable sizes:

One conductor:	**#16 AWG up to 250 kcmil**
Two and three conductors:	**#16 up to #4 AWG**
Four conductors:	**#16 up to #6 AWG**
Seven conductors:	**#16 up to #10 AWG**

Manufactures can supply Type MI cable in sizes **up to 500 kcmil.**

Temperature Ranges

The operating temperature limits of MI-type cable are determined by the fittings and terminations. They may be used **up to 85° Celsius.**

Voltage Ranges

MI cables is intended for use in **600 volt** applications.

Ampacities

Conductors	Size	Amperes
One	16	24
	14	35
	12	40
	10	55
	8	80
	6	105
	4	140
	3	165
	2	190
	1	220
	1/0	260
	2/0	300
	3/0	350
	4/0	405
	250 kcmil	455
Two	16/2	16
	14/2	25
	12/2	32
	10/2	43
	8/2	59
	6/2	79
	4/2	104
Three	16/3	16
	14/3	25
	12/3	32
	10/3	43
	8/3	59
	6/3	79
	4/3	104
Four	16/4	12.8
	14/4	10.0
	12/4	25.6
	10/4	34.4
	8/4	47.2
	6/4	63.2

Seven	16/7	11.2
	14/7	17.5
	12/7	22.4
	10/7	30.1

(Courtesy of Pyrotenax Cable)

Handling and Care

Receiving and Handling.

- Do not allow equipment to touch or damage cable surface or protective wrap.
- If a forklift is used, the forks must be long enough to contact both reel flanges.
- If a crane is used, a shaft through the arbor hold or flange cradles must be employed for the lift.
- If an inclined ramp is used for unloading, it must be wide enough to contact both flanges completely. The reel should be stopped at the bottom by using the flanges, not the cable surface.
- Never drop reels on the ground.

Storage.

- Reels should be stored on a hard, clean surface, not bare earth.
- Store away from open fires or heat sources.
- Cable ends should be resealed as soon as a cable section has been removed.

Receiving Tests. Cable should be tested according to the insulation type and the methods outlined in Chapter 4 of this handbook.

Installation

Securing and Supporting. Type MI cable shall be secured and supported as follows:

- Approved staples, straps, hangers, or similar fittings, as shown in Figure 6–13.
- When mounted on studs, joists, and rafters, Type AC cable shall be installed as shown in Figure 6–14.

- When passed through bored holes, the edge of the hole shall be no less than 1¼ inches from the edge of the stud, joist, or rafter.
- In the event that 1¼ inches cannot be maintained, as mentioned previously and shown in Figure 6–14, a steel spacer or sleeve must be used to protect the cable.
- Mineral insulated cable installed in accessible attics shall be secured, as shown in Figure 6–15.

Termination and Splicing. Figure 6–16 shows a typical termination for mineral insulated cable. Critical items in this termination include the following:

- Exposure of the mineral insulation to air should be minimized.
- The termination should completely seal the insulation from the outside.
- The copper shield should be used as a bonding conductor only when approved by the manufacturer for such service.

Type MI cable does not retain its fire temperature rating when field spliced. The only way to splice MI cable and retain its fire rating is with a factory splice. Note that:

- The splice is actually a double termination with a copper shield placed over it to provide mechanical strength.
- As previously mentioned, the field splice is not rated for the fire temperature rating of the cable itself. Field splices are typically rated at a maximum of 200 to 260° Celsius.

Bending Radius. MI cable shall be bent such that it does not damage the cable. In no case shall the bends be less than those shown in Table 6–3.

Table 6–3 Minimum bending radii for Type MI cable.

Size (External Diameter)	Minimum Bending Radius (Multiple of External Diameter)
Less than ¾ inch (19 mm)	5
Between ¾ inch (19 mm) and 1 inch (25.4 mm)	10

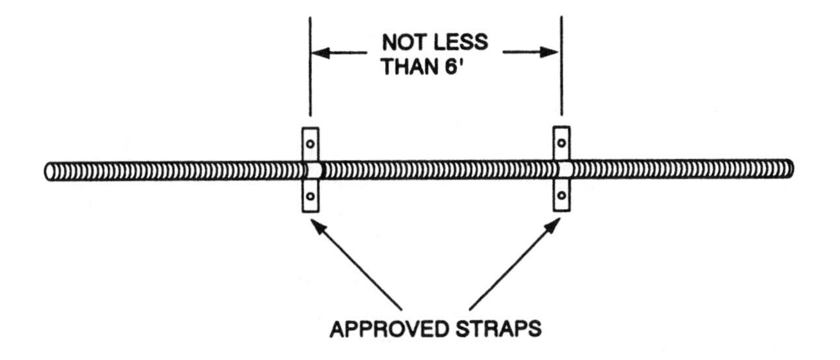

Figure 6–13 Installation of Type MI cable.
(Courtesy of Cadick Corporation)

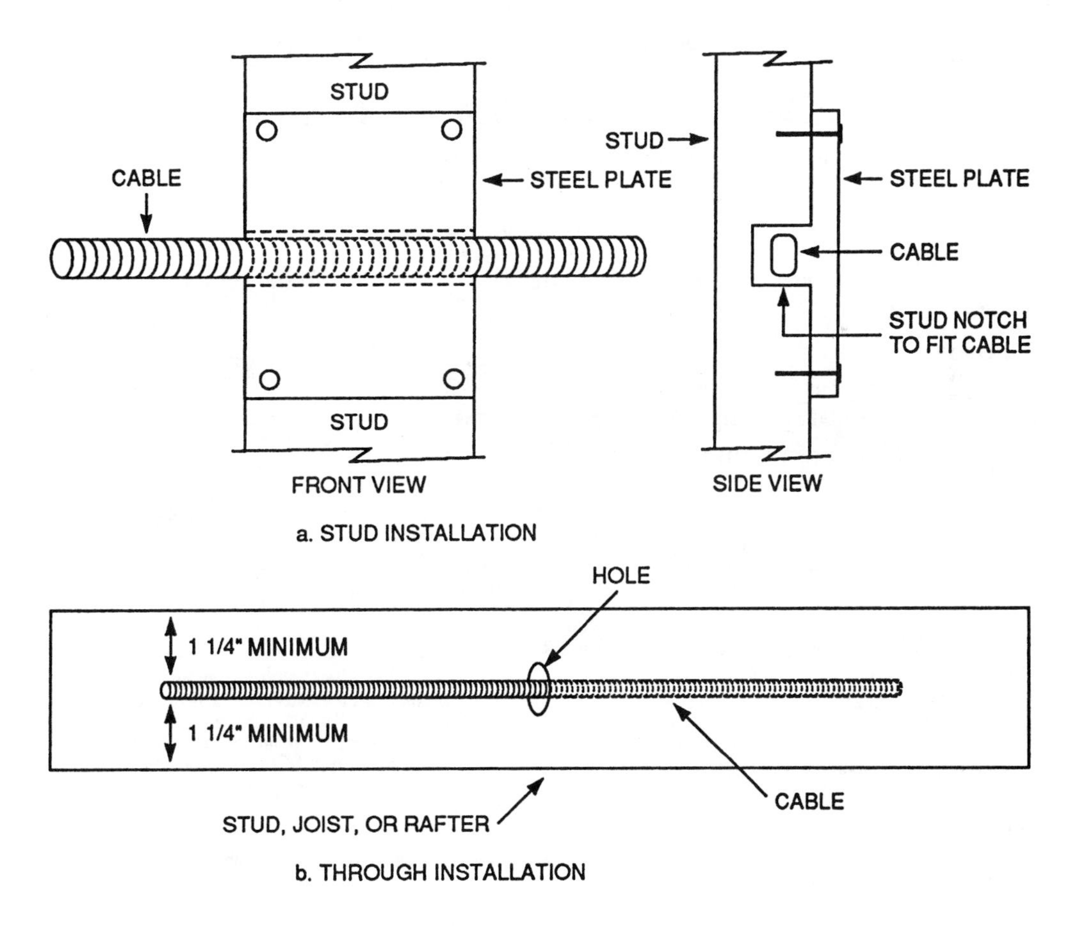

Figure 6–14 Installing cable on studs, joists, and rafters.
(Courtesy of Cadick Corporation)

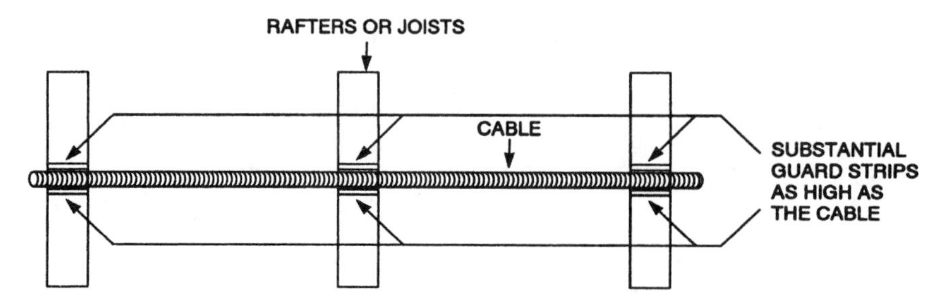

a. OVER JOISTS OR RAFTERS IN ATTICS ACCESSIBLE BY LADDER OR STAIRS,
AS WITHIN 6 FEET OF SCUTTLE HOLE OR ATTIC ENTRANCE

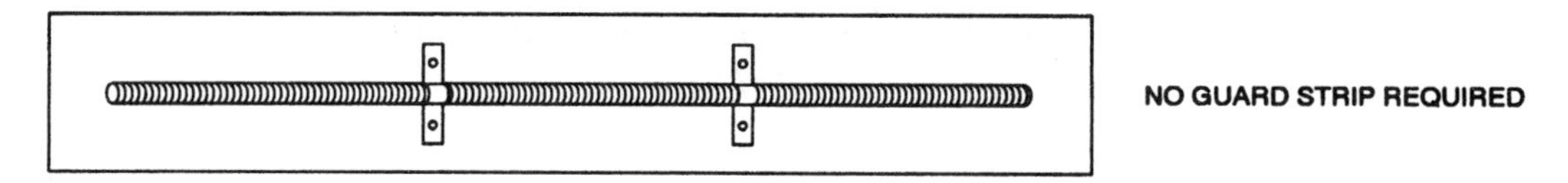

b. ALONG THE SIDE OF JOISTS, STUDS, OR RAFTER

Figure 6–15 Installing cable in accessible attics.
(Courtesy of Cadick Corporation)

TERMINATING 600 VOLT CABLE

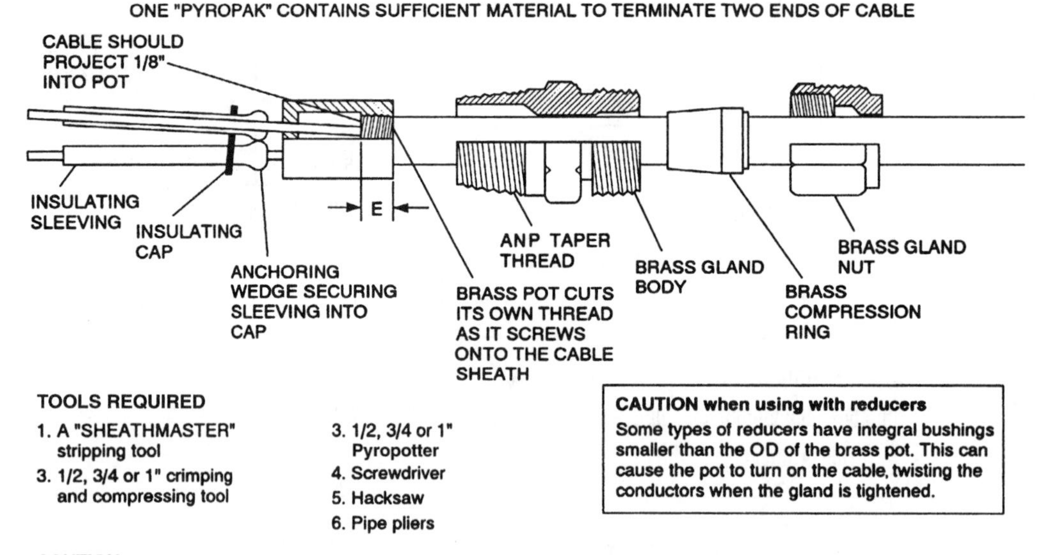

CAUTION

The magnesium oxide insulation in this cable will absorb moisture when exposed to air for any length of time. It is desirable, therefore, to complete a termination once started.

Figure 6–16 Mineral-insulated cable terminations.
(Courtesy of Pyrotenax USA, Inc., East Syracuse, NY)

MACHINE TOOL WIRE (TYPE MTW)

Description

General. Type MTW is made of a single copper conductor surrounded by thermoplastic insulation resistant to moisture, heat, and oil.

Types. N/A.

Applications

Machine tool wire is used in applications where the cable might be exposed to damaging conditions that cannot be minimized or eliminated by other types of systems.

Sizes

Sizes are available from **#22 AWG up to 1,000 kcmil.**

Temperature Ranges

Type MTW is available in both **60 and 90° Celsius** insulation.

Voltage Ranges

Type MTW wire is intended for use in applications **600 volts and lower.**

Ampacities

Type MTW wire ampacities are determined using standard methods from the *NEC®*.

Handling and Care

Receiving and Handling.

- Examine the protective covering for evidence of damage.
- Do not allow equipment to touch or damage cable surface or protective wrap.
- If a forklift is used, the forks must be long enough to contact both reel flanges.

- If a crane is used, a shaft through the arbor hold or flange cradles must be employed for the lift.
- If an inclined ramp is used for unloading, it must be wide enough to contact both flanges completely. The reel should be stopped at the bottom by using the flanges, not the cable surface.
- Never drop reels on the ground.

Storage.

- Reels should be stored on a hard, clean surface, not bare earth.
- Store away from open fires or heat sources.
- Cable ends should be resealed as soon as a cable section has been removed.

Receiving Tests. Cable should be tested according to the insulation type and the methods outlined in Chapter 4 of this handbook.

Installation

Securing and Supporting. MTW wire should be installed and supported as described in NFPA Standard 79. Refer to the latest version of that standard for more specific information.

Termination and Splicing. Type MTW wire is terminated using standard methods, as shown in Figure 6–17. Splicing should be avoided with Type MTW.

Bending Radius. N/A.

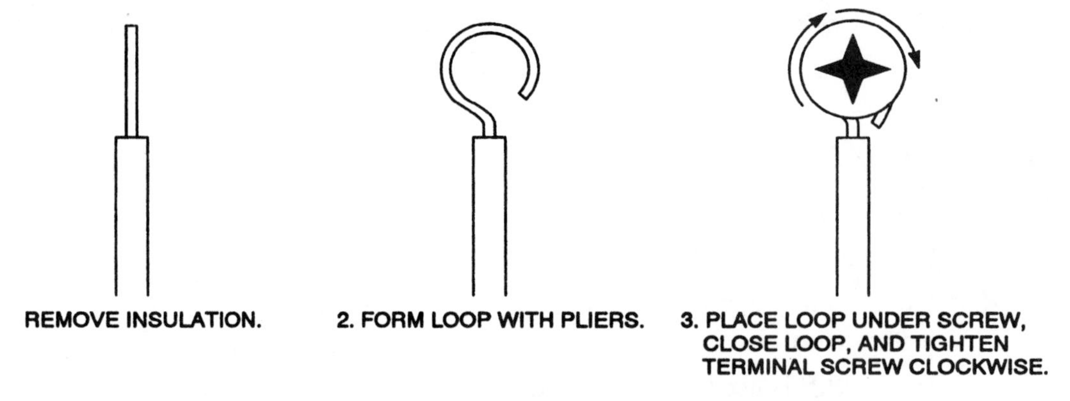

Figure 6–17 Terminating Type MTW cable.

MEDIUM-VOLTAGE CABLE (TYPE MV)

Description

General. Type MV cable is a single- or multiconductor solid dielectric insulated cable rated for 2,001 volts and higher. Figure 6–18 shows a typical MV cable.

Types. N/A.

Applications

MV cables are used on systems up to 35,000 volts as follows:

Permitted	**Not Permitted**
Wet or dry locations	Direct sunlight, unless identified for
Raceways	Cable trays, unless identified for
Cable trays (318)	
Direct burial (710)	
Messenger supported	

NOTE:

Numbers in parentheses indicate articles in the *NEC®* where additional or explanatory information may be found.

Sizes

Type MV cables are available in sizes **up to 4,000 kcmil,** but common applications normally only require sizes up to 750 kcmil.

Temperature Ranges

Type MV cable temperature ranges are determined by the type of insulation used. Standard values **up to 90° Celsius** are used, but high temperature insulations are available.

Voltage Ranges

Type MV cable is available from **2,001 volts up to 35,000 volts.**

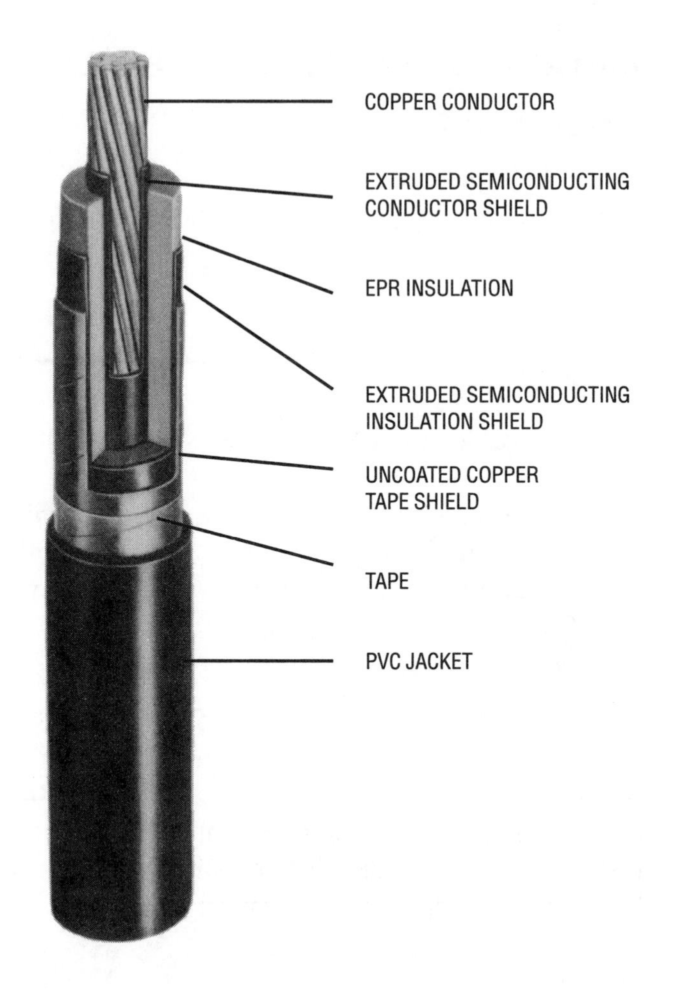

Figure 6–18 Type MV-90 cable.
(Courtesy of General Cable Corporation)

Ampacities

Type MV cable ampacities are determined using standard methods from the *NEC*®.

Handling and Care

Receiving and Handling.

- Examine the protective covering for evidence of damage.
- Do not allow equipment to touch or damage cable surface or protective wrap.

- If a forklift is used, the forks must be long enough to contact both reel flanges.
- If a crane is used, a shaft through the arbor hold or flange cradles must be employed for the lift.
- If an inclined ramp is used for unloading, it must be wide enough to contact both flanges completely. The reel should be stopped at the bottom by using the flanges, not the cable surface.
- Never drop reels on the ground.

Storage.

- Reels should be stored on a hard, clean surface, not bare earth.
- Store away from open fire or heat sources.
- Cable ends should be resealed as soon as a cable section has been removed.

Receiving Tests. Cable should be tested according to the insulation type and the methods outlined in Chapter 4 of this handbook.

Installation

Securing and Supporting. MV cable may be installed above the ground as follows:

- Rigid metal conduit.
- Intermediate metal conduit.
- Rigid nonmetallic conduit.
- Cable trays.
- Busways and cable buses.

MV cable may be direct buried when so identified, and as shown in Figure 6–19.

When direct buried, MV cable must be shielded or be in a continuous metal sheath.

Termination and Splicing. Terminations and splicing of medium-voltage cable is covered in detail in Chapter 3.

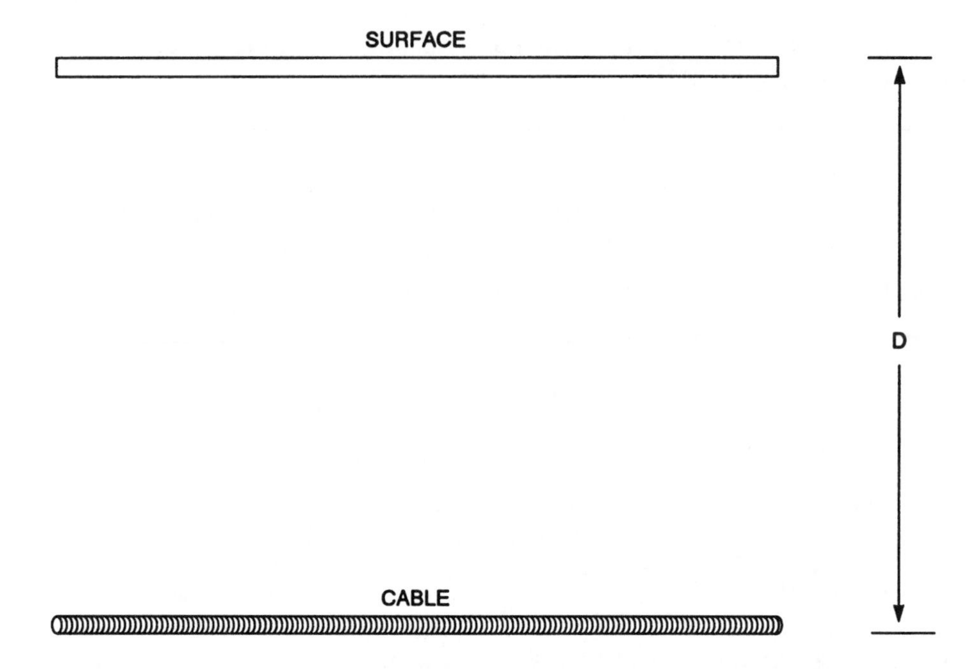

VOLTAGE (KV)	D (MINIMUM, IN INCHES)
BELOW 22	30
22K–40	36
OVER 40	42

Figure 6–19 MV direct burial cable.
(Courtesy of Cadick Corporation)

Bending Radius. Table 6–4 shows the minimum bending radii for various configurations of Type MV cable. Note that the values given are conservative. Refer to manufacturer's specifications for specific cable information.

Table 6–4 Minimum bending radii for MV cable.

Cable Type	Structure or Conditions	Minimum Bending Radius
Single conductors	Unshielded	8 times the overall diameter
	Shielded or lead covered	12 times the overall diameter
Multiconductor or multiplexed single conductors	Select the largest	12 times the diameter of the individual conductors
		7 times the overall diameter

NONMETALLIC-SHEATHED CABLE (TYPES NM, NMC, AND NMS)

Description

General. Nonmetallic-sheathed cable, shown in Figure 6–20, is a factory assembly of two or more insulated conductors having an outer sheath of moisture-resistant, flame-retardant, nonmetallic material.

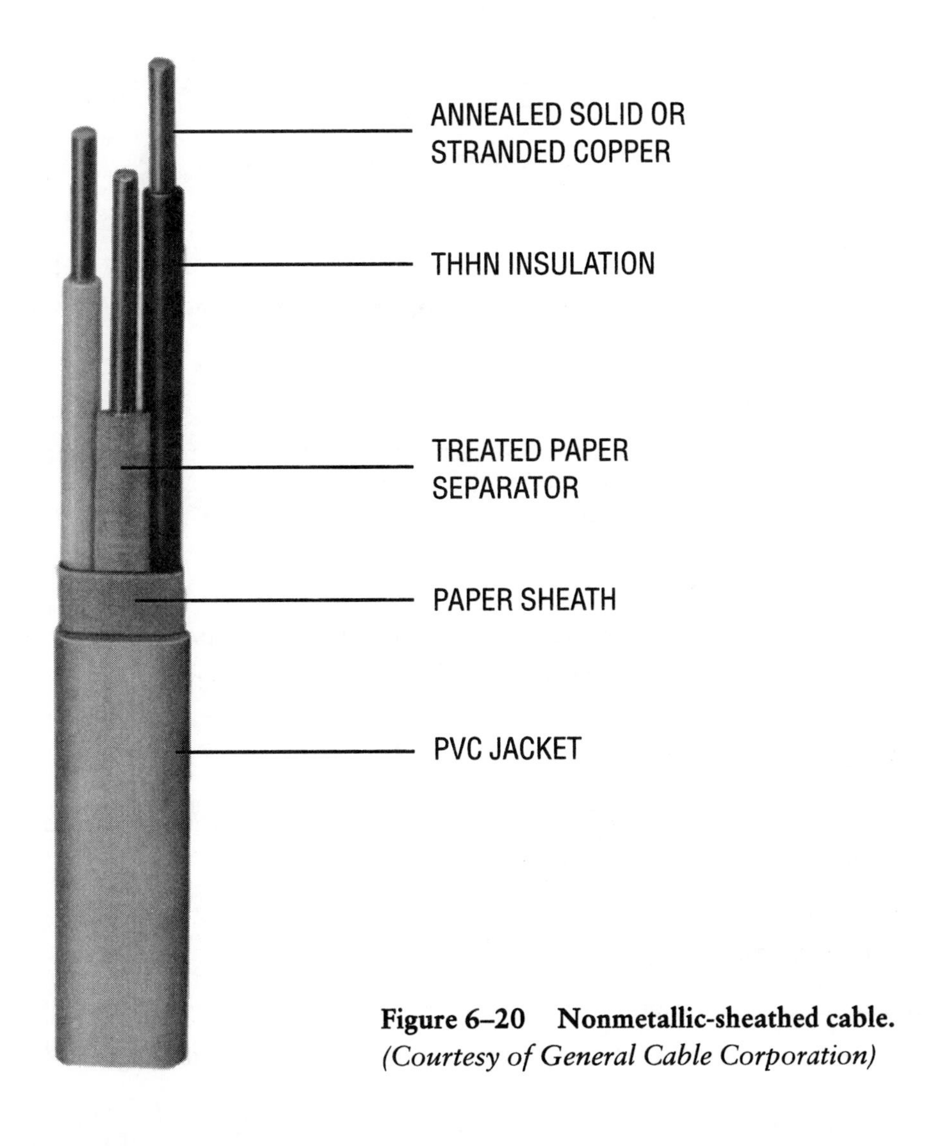

Figure 6–20 Nonmetallic-sheathed cable. *(Courtesy of General Cable Corporation)*

Types. Nonmetallic-sheathed cable is available in the following types:

NM	Normally dry locations.
NMC	Corrosion resistant for dry, moist, damp, or corrosive locations. This type of wire has been given the field name "Romex."
NMS	A factory assembly of insulated power, communications, and signaling conductors enclosed in a common sheath of moisture-resistant, flame-retardant, nonmetallic material. The signal and communications conductors must be separated from the power conductors. Also, the signal conductors may be shielded. Type NMS is intended for use in so-called "smart house" circuits as permitted in Article 780 of the *NEC®*.

Applications

Permitted locations include those listed in Table 6–5.

Sizes

Common sizes available include **#6, #8, #10, #12, and #14 AWG.**

Temperature Ranges

Application temperatures are determined by the insulation type and may be found in the appropriate tables in the *NEC®*.

Voltage Ranges

Nonmetallic sheathed cable is for use in circuits **below 600 volts.** Normal application is for 120/240 volt residential service.

Ampacities

Nonmetallic sheathed cable ampacities are determined using standard methods from the *NEC®*.

Handling and Care

Receiving and Handling.

- Examine the protective covering for evidence of damage.

Table 6–5 Permitted locations for NM, NMC, and NMS cables.

Permitted	NM	• Exposed and concealed work in normally dry locations • May be pulled in air voids in masonry and tile walls if not exposed to excess moisture or dampness
	NMC	• Exposed or concealed work in dry, moist, damp, or corrosive locations • In outside and inside walls of masonry block or tile • In a shallow chase in masonry, concrete, or adobe when properly protected
	NMS	• Exposed and concealed work in normally dry locations • Air voids in masonry block or tile walls when not exposed to excessive dampness
Not Permitted	All Three Types	• Any structure over three stories high • As service entrance cable • In commercial garages with hazardous locations • In theaters, except as provided in *NEC*® Article 518 • In motion picture studios • In storage battery rooms • In hoistways • Embedded in poured concrete • In hazardous locations, except as permitted in the *NEC*®
	Types NM and NMS	• Where exposed to corrosive fumes or vapors • Where embedded in masonry, concrete, or adobe fill or plaster • In shallow chase in masonry, concrete, or adobe and covered with plaster, adobe, or similar finish

- Do not allow equipment to touch or damage cable surface or protective wrap.
- If a forklift is used, the forks must be long enough to contact both reel flanges.
- If a crane is used, a shaft through the arbor hold or flange cradles must be employed for the lift.
- If an inclined ramp is used for unloading, it must be wide enough to contact both flanges completely. The reel should be stopped at the bottom by using the flanges, not the cable surface.
- Never drop reels on the ground.

Storage.

- Reels should be stored on a hard, clean surface, not bare earth.
- Store away from open fires or heat sources.
- Cable ends should be resealed as soon as a cable section has been removed.

Receiving Tests. Cable should be tested according to the insulation type and the methods outlined in Chapter 4 of this handbook.

Installation

Securing and Supporting. Nonmetallic sheathed cable shall be secured and supported as follows:

- Approved staples, straps, hangers, or similar fittings, as shown in Figure 6–21.
- When mounted on studs, joists, and rafters, nonmetallic sheathed cable shall be installed as shown in Figure 6–22.
- When passed through bored holes, the edge of the hole shall be no less than 1¼ inches from the edge of the stud, joist, or rafter.
- In the event that 1¼ inches cannot be maintained, as mentioned previously and shown in Figure 6–3, a steel spacer or sleeve must be used to protect the cable.
- Nonmetallic sheathed cable installed in accessible attics shall be secured, as shown in Figure 6–23.

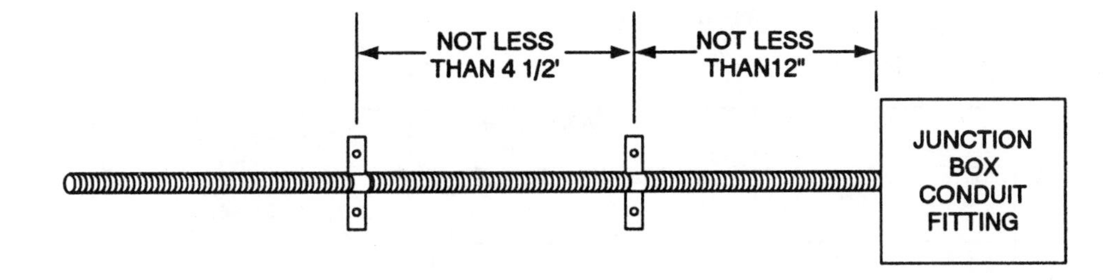

Figure 6–21 Securing nonmetallic-sheathed cable.
(Courtesy of Cadick Corporation)

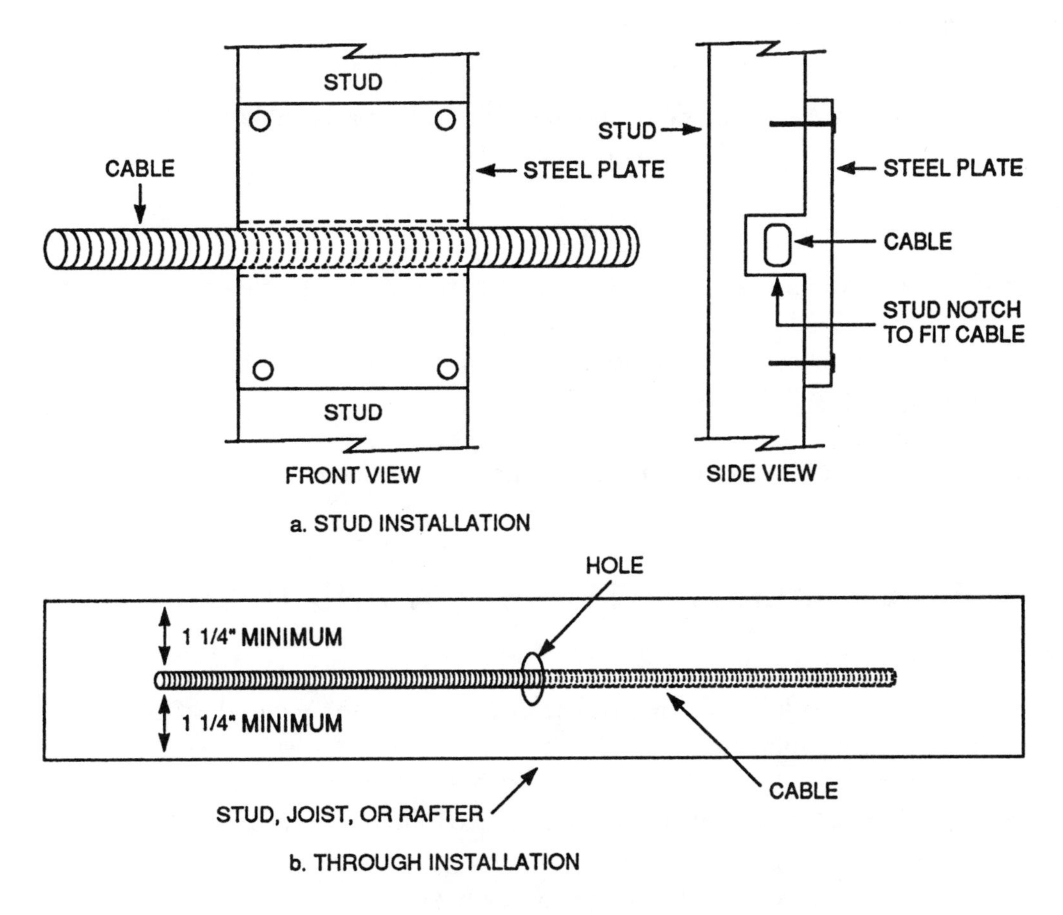

Figure 6–22 Installing cable on studs, joists, and rafters.
(Courtesy of Cadick Corporation)

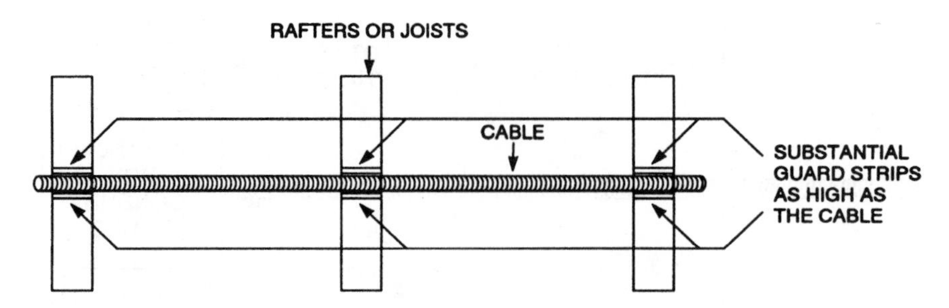

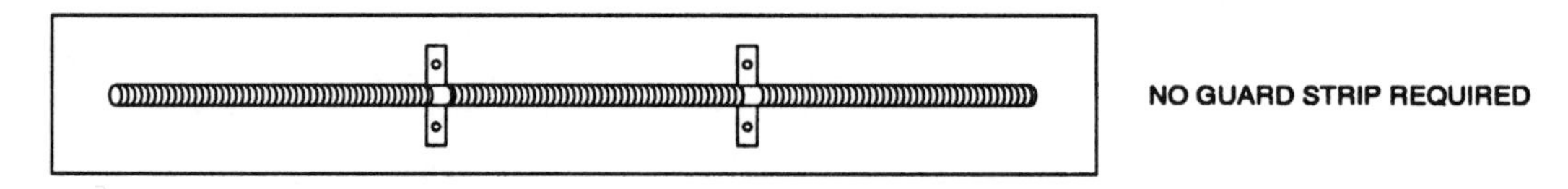

Figure 6–23 Installing nonmetallic cable in accessible attics.
(Courtesy of Cadick Corporation)

Termination and Splicing. Figure 6–24 shows the correct method for terminating nonmetallic sheathed cable.

Nonmetallic sheathed cable splices and terminations must be made in an electrical box approved for such service.

Bending Radius. Not less than five times the cable diameter.

PERFLUOROALKOXY (TYPES PFA AND PFAH)

Description

General. Perfluoroalkoxy is an insulation type used for general wiring and for high-temperature applications.

Types. Perfluoroalkoxy cable is available in the following types:

PFA	Dry locations up to 200° Celsius.
	Dry and wet locations up to 90° Celsius.
PFAH	Dry locations up to 250° Celsius.

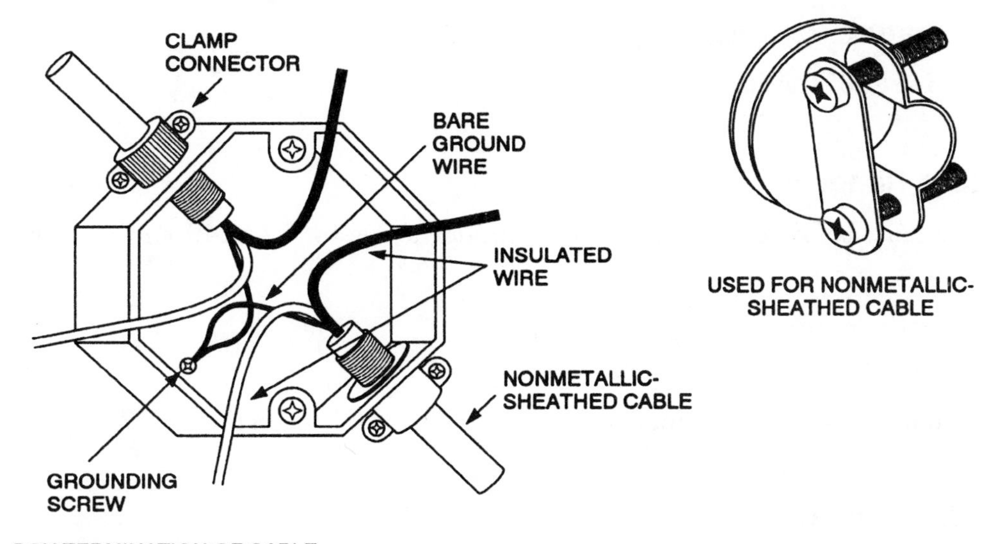

a. BOX TERMINATION OF CABLE

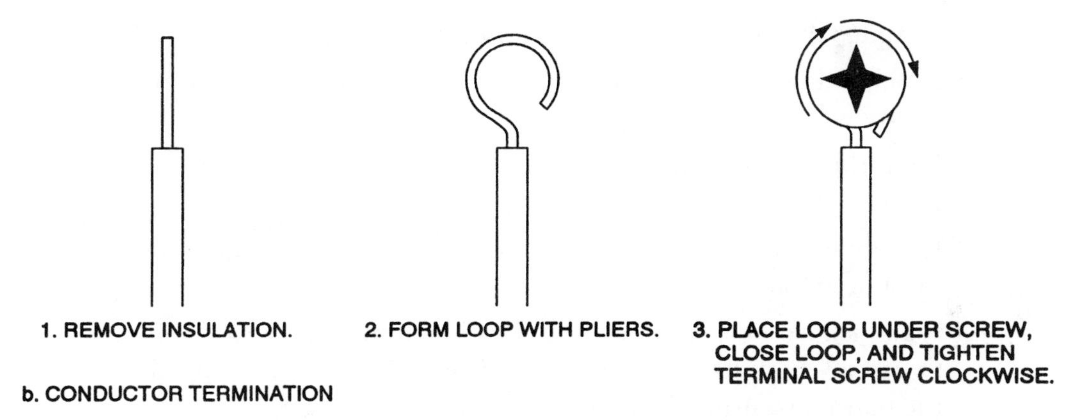

b. CONDUCTOR TERMINATION

Figure 6–24 Terminating nonmetallic-sheathed cables.

Applications

PFA

Dry and damp locations up to 90° Celsius

Dry locations up to 200° Celsius

PFAH

Dry locations up to 250° Celsius

Only for leads within apparatus or within raceways connected to apparatus

Sizes

From **#14 AWG up to 4/0 AWG.**

Temperature Ranges

Up to **250° Celsius**; see applications table.

Voltage Ranges

Circuits **below 600 volts**. Normal application is for 120/240 volt residential service.

Ampacities

Ampacities are determined using standard methods from the *NEC*®.

Handling and Care

Receiving and Handling.

- Examine the protective covering for evidence of damage.
- Do not allow equipment to touch or damage cable surface or protective wrap.
- If a forklift is used, the forks must be long enough to contact both reel flanges.
- If a crane is used, a shaft through the arbor hold or flange cradles must be employed for the lift.
- If an inclined ramp is used for unloading, it must be wide enough to contact both flanges completely. The reel should be stopped at the bottom by using the flanges, not the cable surface.
- Never drop reels on the ground.

Storage.

- Reels should be stored on a hard, clean surface, not bare earth.
- Store away from open fires or heat sources.
- Cable ends should be resealed as soon as a cable section has been removed.

Receiving Tests. Cable should be tested according to the insulation type and the methods outlined in Chapter 4 of this handbook.

Installation

Securing and Supporting. Cable shall be secured and supported as follows:

- Approved staples, straps, hangers, or similar fittings, as shown in Figure 6–21.
- When mounted on studs, joists, and rafters, cable shall be installed as shown in Figure 6–22.
- When passed through bored holes, the edge of the hole shall be no less than 1¼ inches from the edge of the stud, joist, or rafter.
- In the event that 1¼ inches cannot be maintained, as mentioned previously and shown in Figure 6–3, a steel spacer or sleeve must be used to protect the cable.
- Cable installed in accessible attics shall be secured as shown in Figure 6–23.

Termination and Splicing. Figure 6–24 shows the correct method for termination of perfluoroalkoxy cable.

HEAT-RESISTANT RUBBER (TYPES RH, RHH, RHW, AND RHW–2)

Description

General. Heat-resistant rubber is an insulation type that is used for general wiring applications.

Types. Heat-resistant rubber cable is available in the following types:

RH	Dry and damp locations up to 75° Celsius.
RHH	Dry and damp locations up to 90° Celsius.
RHW	Dry and wet locations up to 75° Celsius.
RHW–2	Dry and wet locations up to 90° Celsius.

Applications

Used for all general wiring requirements for low- and medium-voltage applications.

Sizes

From **#14 AWG up to 2,000 kcmil.**

Temperature Ranges

Up to 90° Celsius.

Voltage Ranges

Circuits **below 600 volts up to 35,000 volts.**

Ampacities

Ampacities are determined using standard methods from the *NEC*®.

Handling and Care

Receiving and Handling.

- Examine the protective covering for evidence of damage.
- Do not allow equipment to touch or damage cable surface or protective wrap.
- If a forklift is used, the forks must be long enough to contact both reel flanges.
- If a crane is used, a shaft through the arbor hold or flange cradles must be employed for the lift.
- If an inclined ramp is used for unloading, it must be wide enough to contact both flanges completely. The reel should be stopped at the bottom by using the flanges, not the cable surface.
- Never drop reels on the ground.

Storage.

- Reels should be stored on a hard, clean surface, not bare earth.
- Store away from open fires or heat sources.
- Cable ends should be resealed as soon as a cable section has been removed.

Receiving Tests. Cable should be tested according to the insulation type and the methods outlined in Chapter 4 of this handbook.

Installation

Securing and Supporting. Low-voltage cable shall be secured and supported as follows:

- Approved staples, straps, hangers, or similar fittings, as shown in Figure 6–21.
- When mounted on studs, joists, and rafters, cable shall be installed as shown in Figure 6–22.
- When passed through bored holes, the edges of the hole shall be no less than 1¼ inches from the edge of the stud, joist, or rafter.
- In the event that 1¼ inches cannot be maintained, as mentioned previously and shown in Figure 6–3, a steel spacer or sleeve must be used to protect the cable.
- Cable installed in accessible attics shall be secured as shown in Figure 6–23.

Medium-voltage cable should be installed as described previously for Type MV cable.

Termination and Splicing. Terminations depend on application voltages. See the appropriate sections for complete information.

SILICONE ASBESTOS (TYPE SA)

Description

General. Silicone asbestos (Type SA) features a silicone insulation layer covered by a heat-resistant jacket. Early versions of this cable used asbestos for the jacket. Concerns about asbestos have caused asbestos to be replaced by glass and aramid fiber, however. Type SA wire provides relatively high ampacities and good temperature performance.

Types. N/A.

Applications

Used for general wiring applications in high ambient temperature locations, including power plants, boilers, and ovens.

Sizes

From **#14 AWG up to 2,000 kcmil.**

Temperature Ranges

Up to 90° Celsius in dry and damp locations.
Up to 125° Celsius in special applications.

Voltage Ranges

Circuits **below 600 volts.**

Ampacities

Ampacities are determined using standard methods from the *NEC®*.

Handling and Care

Receiving and Handling.

- Examine the protective covering for evidence of damage.
- Do not allow equipment to touch or damage cable surface or protective wrap.
- If a forklift is used, the forks must be long enough to contact both reel flanges.
- If a crane is used, a shaft through the arbor hold or flange cradles must be employed for the lift.
- If an inclined ramp is used for unloading, it must be wide enough to contact both flanges completely. The reel should be stopped at the bottom by using the flanges, not the cable surface.
- Never drop reels on the ground.

Storage.

- Reels should be stored on a hard, clean surface, not bare earth.

- Store away from open fires or heat sources.
- Cable ends should be resealed as soon as a cable section has been removed.

Receiving Tests. Cable should be tested according to the insulation type and the methods outlined in Chapter 4 of this handbook.

Installation

Securing and Supporting. Cable shall be secured and supported as follows:

- Approved staples, straps, hangers, or similar fittings, as shown in Figure 6–21.
- When mounted on studs, joists, and rafters, cable shall be installed as shown in Figure 6–22.
- When passed through bored holes, the edge of the hole shall be no less than 1¼ inches from the edge of the stud, joist, or rafter.
- In the event that 1¼ inches cannot be maintained, as mentioned previously and shown in Figure 6–3, a steel spacer or sleeve must be used to protect the cable.
- Cable installed in accessible attics shall be secured as shown in Figure 6–23.

Termination and Splicing. Terminations depend on application voltages. See the appropriate sections for complete information.

SERVICE ENTRANCE CABLE (TYPES SE AND USE)

Description

General. Service entrance cable is a single- or multiconductor assembly provided with or without an overall covering and is primarily used in service entrances. Conductors used may be Type RH, RHW, RHH, or XHHW (see Figure 6–25).

Types. Service entrance cable is available in the following types:

SE	Cable with a flame-retardant, moisture-resistant covering.
USE	Cable for underground service use. It has a moisture-resistant covering and sometimes a flame-retardant covering also.
USE-2	Same as Type USE but with higher rated insulation.

Figure 6–25 USE-2 cable.
(Courtesy of Rome Cable Corporation)

Applications

- Service entrance conductors; Article 230 of the *NEC®*.
- Branch circuits or feeders with insulated ground conductor.

Sizes

Sizes **up to 2,000 kcmil** are available.

Temperature Ranges

Types SE and USE: **75° Celsius.**
Type USE-2: **90° Celsius.**

Voltage Ranges

Intended for applications **600 volts and lower.**

Ampacities

Service entrance cable ampacities are determined using standard methods from the *NEC®*.

Handling and Care

Receiving and Handling.

- Examine the protective covering for evidence of damage.
- Do not allow equipment to touch or damage cable surface or protective wrap.

- If a forklift is used, the forks must be long enough to contact both reel flanges.
- If a crane is used, a shaft through the arbor hold or flange cradles must be employed for the lift.
- If an inclined ramp is used for unloading, it must be wide enough to contact both flanges completely. The reel should be stopped at the bottom by using the flanges, not the cable surface.
- Never drop reels on the ground.

Storage.

- Reels should be stored on a hard, clean surface, not bare earth.
- Store away from open fires or heat sources.
- Cable ends should be resealed as soon as a cable section has been removed.

Receiving Tests. Cable should be tested according to the insulation type and the methods outlined in Chapter 4 of this handbook.

Installation

Securing and Supporting. When used in interior wiring, service entrance cable shall be secured and supported as follows:

- Approved staples, straps, hangers, or similar fittings, as shown in Figure 6–2.
- When mounted on studs, joists, and rafters, Type SE and USE cables shall be installed, as shown in Figure 6–3.
- When passed through bored holes, the edge of the hole shall be no less than 1¼ inches from the edge of the stud, joist, or rafter.
- In the event that 1¼ inches cannot be maintained, as mentioned previously and shown in Figure 6–3, a steel spacer or sleeve must be used to protect the cable.
- Service entrance cable installed in accessible attics shall be secured as shown in Figure 6–4.

When installed as service entrance, cable installation shall comply with Article 230 of the *NEC*®.

Termination and Splicing. Termination of service entrance cable shall be made according to the procedures discussed in Chapter 3. Splices and termination shall be made in a box or fitting approved for such service.

Bending Radius. Not less than five times the cable diameter.

SYNTHETIC HEAT RESISTANT (TYPE SIS)

Description

General. Synthetic heat-resistant wire (Type SIS) is a single-conductor copper wire insulated with heat-resistant rubber. It is intended for use in wiring switchboards and other control panel applications. The conductor of Type SIS may be solid or stranded.

Types. N/A.

Applications

For switchboard wiring only.

Sizes

From **#14 AWG up to 4/0 AWG.**

Temperature Ranges

Up to 90° Celsius.

Voltage Ranges

Circuits **below 600 volts.** Normal application is in low-voltage control applications.

Ampacities

Ampacities are determined using standard methods from the *NEC®*.

Handling and Care

Receiving and Handling.

- Examine the protective covering for evidence of damage.
- Do not allow equipment to touch or damage cable surface or protective wrap.
- If a forklift is used, the forks must be long enough to contact both reel flanges.
- If a crane is used, a shaft through the arbor hold or flange cradles must be employed for the lift.
- If an inclined ramp is used for unloading, it must be wide enough to contact both flanges completely. The reel should be stopped at the bottom by using the flanges, not the cable surface.
- Never drop reels on the ground.

Storage.

- Reels should be stored on a hard, clean surface, not bare earth.
- Store away from open fires or heat sources.
- Cable ends should be resealed as soon as a cable section has been removed.

Receiving Tests. Cable should be tested according to the insulation type and the methods outlined in Chapter 4 of this handbook.

Installation

Securing and Supporting. The installation of Type SIS is determined by the particular switchboard application. Cable clamps, cable ties, and lacing twine are all acceptable methods.

Termination and Splicing. Type SIS may be terminated using squeeze-on lugs. Smaller sizes may be terminated directly using methods shown in Figure 6–17.

THERMOPLASTIC ASBESTOS (TYPE TA)

Description

General. Thermoplastic asbestos is an insulation system that features a thermoplastic insulation layer covered by a heat-resistant jacket. Asbestos has been replaced by glass or aramid fiber jackets. Type TA wire provides relatively high ampacities, good temperature performance, and relatively high resistance to abrasion.

Types. N/A.

Applications

Used for switchboard applications.

Sizes

From **#14 AWG up to 4/0 AWG.**

Temperature Ranges

Up to 90° Celsius.

Voltage Ranges

Circuits **below 600 volts.**

Ampacities

Ampacities are determined using standard methods from the *NEC*®.

Handling and Care

Receiving and Handling.

- Examine the protective covering for evidence of damage.
- Do not allow equipment to touch or damage cable surface or protective wrap.
- If a forklift is used, the forks must be long enough to contact both reel flanges.

- If a crane is used, a shaft through the arbor hold or flange cradles must be employed for the lift.
- If an inclined tramp is used for unloading, it must be wide enough to contact both flanges completely. The reel should be stopped at the bottom by using the flanges, not the cable surface.
- Never drop reels on the ground.

Storage.

- Reels should be stored on a hard, clean surface, not bare earth.
- Store away from open fires or heat sources.
- Cable ends should be resealed as soon as a cable section has been removed.

Receiving Tests. Cable should be tested according to the insulation type and the methods outlined in Chapter 4 of this handbook.

Installation

Securing and Supporting. The installation of Type TA wire is determined by the particular switchboard application. Cable clamps, cable ties, and lacing twine are all acceptable methods.

Termination and Splicing. Type TA wire may be terminated using squeeze-on lugs. Smaller sizes may be terminated directly using methods down in Figure 6–17.

THERMOPLASTIC AND FIBROUS OUTER BRAID (TYPE TBS)

Description

General. Thermoplastic and fibrous outer braid is an insulation system that features a thermoplastic insulation layer covered by a fibrous jacket. Type TBS wire provides relatively high ampacities, good temperature performance, and relatively high resistance to abrasion.

Types. N/A.

Applications

Used for switchboard applications.

Sizes

From **#14 AWG up to 4/0 AWG.**

Temperature Ranges

Up to 90° Celsius.

Voltage Ranges

Circuits **below 600 volts.**

Ampacities

Ampacities are determined using standard methods from the *NEC*®.

Handling and Care

Receiving and Handling.

- Examine the protective covering for evidence of damage.
- Do not allow equipment to touch or damage cable surface or protective wrap.
- If a forklift is used, the forks must be long enough to contact both reel flanges.
- If a crane is used, a shaft through the arbor hold or flange cradles must be employed for the lift.
- If an inclined ramp is used for unloading, it must be wide enough to contact both flanges completely. The reel should be stopped at the bottom by using the flanges, not the cable surface.
- Never drop reels on the ground.

Storage.

- Reels should be stored on a hard, clean surface, not bare earth.
- Store away from open fires or heat sources.

- Cable ends should be resealed as soon as a cable section has been removed.

Receiving Tests. Cable should be tested according to the insulation type and the methods outlined in Chapter 4 of this handbook.

Installation

Securing and Supporting. The installation of Type TBS wire is determined by the particular switchboard application. Cable clamps, cable ties, and lacing twine are all acceptable methods.

Termination and Splicing. Type TBS wire may be terminated using squeeze-on lugs. Smaller sizes may be terminated directly using methods shown in Figure 6–17.

TRAY CABLE (TYPE TC)

Description

General. Type TC power and control cable is a factory assembly of two or more insulated conductors, with or without bare or covered grounding conductors under a nonmetallic sheath. This type of cable is approved for installation in cable trays, raceways, or where supported by a messenger wire. Figure 6–26 is a typical Type TC cable.

Types. N/A.

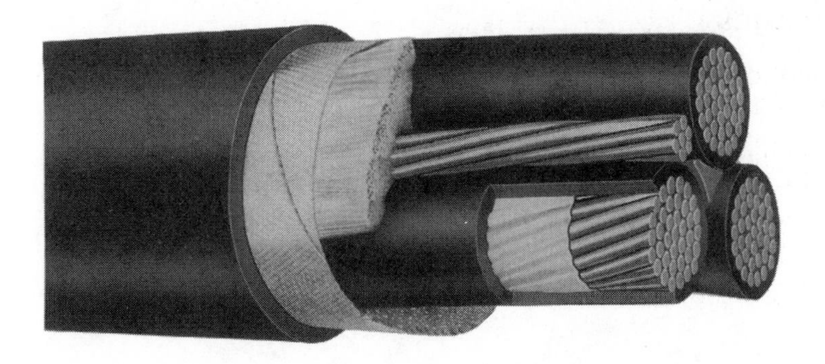

Figure 6–26 Type TC cable.
(Courtesy of Rome Cable Corporation)

Applications

Permitted	Not Permitted
Power, lighting, and control	Where exposed to physical damage
Cable trays	Open cable on brackets or cleats
Raceways	Exposed to UV, unless identified
Messenger wire supported	Direct buried, unless identified
Hazardous locations (318,501,502,504)	
Class 1 Circuits (725)	

NOTE:

Numbers in parentheses indicate articles in the *NEC®* where additional or explanatory information may be found.

Sizes

Copper	Aluminum or Copper-coated aluminum
#18 AWG up to 1,000 kcmil#	**12 AWG up to 1,000 kcmil**

Temperature Ranges

The temperature range of Type TC cable is determined by the insulation used on the individual conductors. Temperature ranges may be found in the *NEC®*.

Voltage Ranges

Intended for application at **600 volts and below.**

Ampacities

Type TC cable ampacities for conductors larger than #14 AWG are determined using standard methods from the *NEC®*.

Handling and Care

Receiving and Handling.

- Examine the protective covering for evidence of damage.
- Do not allow equipment to touch or damage cable surface or protective wrap.
- If a forklift is used, the forks must be long enough to contact both reel flanges.
- If a crane is used, a shaft through the arbor hold or flange cradles must be employed for the lift.
- If an inclined ramp is used for unloading, it must be wide enough to contact both flanges completely. The reel should be stopped at the bottom by using the flanges, not the cable surface.
- Never drop reels on the ground.

Storage.

- Reels should be stored on a hard, clean surface, not bare earth.
- Store away from open fires or heat sources.
- Cable ends should be resealed as soon as a cable section has been removed.

Receiving Tests. Cable should be tested according to the insulation type and the methods outlined in Chapter 4 of this handbook.

Installation

Securing and Supporting. Tray cable may only be installed in trays, raceways, or by the method described previously. *NEC*® Articles 300 and 318 provide installation details.

Termination and Splicing. Termination of tray cable depends on the type of insulation used in the cable. Splices and terminations are only allowed in the tray using approved boxes or fittings for such an application.

Bending Radius. Bends shall be made so as to not damage the cable.

EXTENDED POLYTETRAFLUOROETHYLENE (TYPE TFE)

Description

General. Type TFE is a special high-temperature insulation. It is used for leads connected to apparatus or within raceways connected to apparatus.

Types. N/A.

Applications

- Dry locations only.
- Leads within apparatus or in raceways connected to apparatus.
- Open wiring.

Sizes

From **#14 AWG up to 4/0 AWG.**

Temperature Ranges

Up to **250° Celsius.**

Voltage Ranges

Circuits **below 600 volts.**

Ampacities

Ampacities are determined using standard methods from the *NEC®*.

Handling and Care

Receiving and Handling.

- Examine the protective covering for evidence of damage.
- Do not allow equipment to touch or damage cable surface or protective wrap.
- If a forklift is used, the forks must be long enough to contact both reel flanges.

- If a crane is used, a shaft through the arbor hold or flange cradles must be employed for the lift.
- If an inclined ramp is used for unloading, it must be wide enough to contact both flanges completely. The reel should be stopped at the bottom by using the flanges, not the cable surface.
- Never drop reels on the ground.

Storage.

- Reels should be stored on a hard, clean surface, not bare earth.
- Store away from open fires or heat sources.
- Cable ends should be resealed as soon as a cable section has been removed.

Receiving Tests. Cable should be tested according to the insulation type and the methods outlined in Chapter 4 of this handbook.

Installation

Securing and Supporting. Cable shall be secured and supported as follows:

- Approved staples, straps, hangers, or similar fittings, as shown in Figure 6–21.
- When mounted on studs, joists, and rafters, cable shall be installed as shown in Figure 6–22.
- When passed through bored holes, the edge of the hole shall be no less than 1¼ inches from the edge of the stud, joist, or rafter.
- In the event that 1¼ inches cannot be maintained, as mentioned previously and shown in Figure 6–3, a steel spacer or sleeve must be used to protect the cable.
- Cable installed in accessible attics shall be secured as shown in Figure 6–23.

Termination and Splicing. Figure 6–24 shows the correct methods for termination of Type TFE.

THERMOPLASTIC (TYPES THHN, THHW, THW, THWN, AND TW)

Description

General. Thermoplastic insulations are modern insulations used in many power and control installations.

Types. Heat-resistant thermoplastic cable is available in the following types:

THHN Flame retardant, heat resistant, with nylon jacket.
THHW Flame retardant, moisture and heat resistant.
THW Flame retardant, moisture and heat resistant.
THWN Flame retardant, moisture and heat resistant, with nylon jacket.
TW Flame retardant, moisture resistant.

Applications

Used for all general wiring requirements for low- and medium-voltage applications.

Sizes

Types THW, TW:	From **#14 AWG up to 2,000 kcmil.**
Types THHN, THHW, THWN:	From **#14 AWG up to 1,000 kcmil.**

Temperature Ranges

Up to 90° Celsius.

Voltage Ranges

Circuits **below 600 volts up to 35,000 volts.**

Ampacities

Ampacities are determined using standard methods from the *NEC®*.

Handling and Care

Receiving and Handling.

- Examine the protective covering for evidence of damage.
- Do not allow equipment to touch or damage cable surface or protective wrap.
- If a forklift is used, the forks must be long enough to contact both reel flanges.
- If a crane is used, a shaft through the arbor hold or flange cradles must be employed for the lift.
- If an inclined ramp is used for unloading, it must be wide enough to contact both flanges completely. The reel should be stopped at the bottom by using the flanges, not the cable surface.
- Never drop reels on the ground.

Storage.

- Reels should be stored on a hard, clean surface, not bare earth.
- Store away from open fires or heat sources.
- Cable ends should be resealed as soon as a cable section has been removed.

Receiving Test. Cable should be tested according to the insulation type and the methods outlined in Chapter 4 of this handbook.

Installation

Securing and Supporting. Low-voltage cable shall be secured and supported as follows:

- Approved staples, straps, hangers, or similar fittings, as shown in Figure 6–21.
- When mounted on studs, joists, and rafters, cable shall be installed as shown in Figure 6–22.
- When passed through bored holes, the edge of the hole shall be no less than 1¼ inches from the edge of the stud, joist, or rafter.
- In the event that 1¼ inches cannot be maintained, as mentioned previously and in Figure 6–3, a steel spacer or sleeve must be used to protect the cable.

- Cable installed in accessible attics shall be secured as shown in Figure 6–23.

High-voltage cable should be installed as described previously for Type MV cable.

Termination and Splicing. Terminations depend on application voltages. See the appropriate sections for complete information.

UNDERGROUND FEEDER AND BRANCH CIRCUIT CABLE (TYPE UF)

Description

General. Type UF is a fabricated assembly of one or more individually insulated conductors. Insulation used must be of a moisture-resistant type.

Types. N/A.

Applications

Permitted	Not Permitted
Underground direct burial	Service entrance cable
Interior wiring in all locations	Commercial garages
	Theaters
	Motion picture studios
	Storage battery rooms
	Hoistways
	Hazardous locations
	Embedded in poured concrete (424)
	Exposed to sunlight, unless identified

NOTE:

Numbers in parentheses indicate articles in the *NEC®* where additional or explanatory information may be found.

Sizes

Copper	Aluminum or Copper-coated aluminum
#14 AWG up to 4/0 AWG	#12 AWG up to 4/0 AWG

Temperature Ranges

60° Celsius.

Voltage Ranges

600 volts and lower.

Ampacities

Type UF cable ampacities are determined using the 60° Celsius conductor entries from the current edition of the *NEC®*.

Handling and Care

Receiving and Handling.

- Examine the protective covering for evidence of damage.
- Do not allow equipment to touch or damage cable surface or protective wrap.
- If a forklift is used, the forks must be long enough to contact both reel flanges.
- If a crane is used, a shaft through the arbor hold or flange cradles must be employed for the lift.
- If an inclined ramp is used for unloading, it must be wide enough to contact both flanges completely. The reel should be stopped at the bottom by using the flanges, not the cable surface.
- Never drop reels on the ground.

Storage.

- Unjacketed, metal-clad cable should be stored indoors to avoid corrosion.
- Reels should be stored on a hard, clean surface, not bare earth.

- Store away from open fires or heat sources.
- Cable ends should be resealed as soon as a cable section has been removed.

Receiving Tests. Cable should be tested according to the insulation type and the methods outlined in Chapter 4 of this handbook.

Installation

Securing and Supporting. When used for interior wiring, Type UF cable shall be secured and supported as follows:

- Approved staples, straps, hangers, or similar fittings, as shown in Figure 6–10.
- In cable trays where approved for that application.
- Direct buried, as per Figure 6–11.
- Type UF cable may be used as overhead or underground service entrance cable.

Termination and Splicing. Type UF cable shall be terminated and spliced according to the low-voltage methods presented in Chapter 3 of this handbook. Underground splices shall be made using methods that render the cable impervious to moisture.

Bending Radius. Bends shall be made so as not to damage the cable.

VARNISHED CAMBRIC (TYPE V)

Description

General. Varnished cambric is an insulated cable composed of conductors wrapped with cotton tape that has been coated with insulating varnish. The layers of tape are separated with a layer of nonhardening mineral compound. The mineral compound serves as a lubricant.

If the cable is to be used in a wet location, it *must* be covered with a moisture-proof jacket, such as lead or thermoplastic.

Varnished cambric is an older insulation whose electrical characteristics fall somewhere between paper and thermoplastic.

Types. N/A.

Applications

Used for all general wiring requirements for low- and medium-voltage applications in dry locations only.

Sizes

From **#14 AWG up to 2,000 kcmil.** Sizes smaller than #6 AWG are by special permission only.

Temperature Ranges

Up to **85° Celsius.**

Voltage Ranges

Circuits **below 600 volts up to 35,000 volts.**

Ampacities

Ampacities are determined using standard methods from the *NEC*®.

Handling and Care

Receiving and Handling.

- Examine the protective covering for evidence of damage.
- Do not allow equipment to touch or damage cable surface or protective wrap.
- If a forklift is used, the forks must be long enough to contact both reel flanges.
- If a crane is used, a shaft through the arbor hold or flange cradles must be employed for the lift.
- If an inclined ramp is used for unloading, it must be wide enough to contact both flanges completely. The reel should be stopped at the bottom by using the flanges, not the cable surface.
- Never drop reels on the ground.

Storage.

- Reels should be stored on a hard, clean surface, not bare earth.
- Store away from open fires or heat sources.
- Cable ends should be resealed as soon as a cable section has been removed.
- Type V cable should never be stored in a moist or wet area.

Receiving Tests. Cable should be tested according to the insulation type and the methods outlined in Chapter 4 of this handbook.

Installation

Securing and Supporting. Low-voltage cable shall be secured and supported as follows:

- Approved staples, straps, hangers, or similar fittings, as shown in Figure 6–21.
- When mounted on studs, joints, and rafters, cable shall be installed as shown in Figure 6–22.
- When passed through bored holes, the edge of the hole shall be no less than 1¼ inches from the edge of the stud, joist, or rafter.
- In the event that 1¼ inches cannot be maintained, as mentioned previously and in Figure 6–3, a steel spacer or sleeve must be used to protect the cable.
- Cable installed in accessible attics shall be secured as shown in Figure 6–23.

High-voltage cable should be installed as described previously for Type MV cable.

Termination and Splicing. Terminations depend on application voltages. See the appropriate sections for complete information.

CROSS-LINKED POLYMER (TYPES XHHW AND XHHW-2)

Description

General. An insulation made of a cross-linked polymer, such as polyethylene. These types of thermosetting insulations have superior temperature characteristics.

Types. Types XHHW and XHHW-2 are both cross-linked polymer types. Type XHHW-2 is a newer listing and is allowed in wet or dry installations to 90° Celsius.

Applications

Used for all general wiring requirements for low- and medium-voltage applications.

Sizes

From **#14 AWG up to 2,000 kcmil.**

Temperature Ranges

Type XHHW:	Dry locations **up to 90° Celsius.** Wet locations **up to 75° Celsius.**
Type XHHW-2:	Wet or dry locations **up to 90° Celsius.**

Voltage Ranges

Circuits **below 600 volts up to 35,000 volts.**

Ampacities

Ampacities are determined using standard methods from the *NEC*®.

Handling and Care

Receiving and Handling.

- Examine the protective covering for evidence of damage.
- Do not allow equipment to touch or damage cable surface or protective wrap.

- If a forklift is used, the forks must be long enough to contact both reel flanges.
- If a crane is used, a shaft through the arbor hold or flange cradles must be employed for the lift.
- If an inclined ramp is used for unloading, it must be wide enough to contact both flanges completely. The reel should be stopped at the bottom by using the flanges, not the cable surface.
- Never drop reels on the ground.

Storage.

- Reels should be stored on a hard, clean surface, not bare earth.
- Store away from open fires or heat sources.
- Cable ends should be resealed as soon as a cable section has been removed.

Receiving Tests. Cable should be tested according to the insulation type and the methods outlined in Chapter 4 of this handbook.

Installation

Securing and Supporting. Low-voltage cable shall be secured and supported as follows:

- Approved staples, straps, hangers, or similar fittings, as shown in Figure 6–21.
- When mounted on studs, joists, and rafters, cable shall be installed as shown in Figure 6–22.
- When passed through bored holes, the edge of the hole shall be no less than 1¼ inches from the edge of the stud, joist, or rafter.
- In the event that 1¼ inches cannot be maintained, as mentioned previously and shown in Figure 6–3, a steel spacer or sleeve must be used to protect the cable.
- Cable installed in accessible attics shall be secured as shown in Figure 6–23.

High-voltage cable should be installed as described previously for Type MV cable.

Termination and Splicing. Terminations depend on application voltages. See the appropriate sections for complete information.

MODIFIED ETHYLENE TETRAFLUOROETHYLENE (TYPES Z AND ZW)

Description

General. Types Z and ZW are insulation systems that have excellent temperature and radiation characteristics. Figure 6–27 shows a typical Type Z cable.

Types. Modified ethylene tetrafluoroethylene is available in the following types:

Z	Dry and damp locations. Dry locations; special applications
ZW	Wet locations. Dry and damp locations. Dry locations; special applications.

Applications

Type Z cables are used in areas such as nuclear power plants and fossil fuel plants where ruggedness, high temperature, and radiation resistance are important.

Sizes

Type Z: From **#14 AWG up to 4/0 AWG.**
Type ZW: From **#14 AWG up to #2 AWG.**

Temperature Ranges

Type Z: **Up to 90° Celsius** for dry and damp locations.
Up to 150° Celsius for dry locations and special applications.

Figure 6–27 Type Z wire.
(Courtesy of The Okonite Company—Power, Control, and Instrumentation Cables)

Type ZW: **Up to 75° Celsius** for wet locations.
Up to 90° Celsius for dry and damp locations.
Up to 150° Celsius for dry locations and special applications.

Voltage Ranges

Circuits **below 600 volts.**

Ampacities

Ampacities are determined using standard methods from the *NEC*®.

Handling and Care

Receiving and Handling.

- Examine the protective covering for evidence of damage.
- Do not allow equipment to touch or damage cable surface or protective wrap.
- If a forklift is used, the forks must be long enough to contact both reel flanges.
- If a crane is used, a shaft through the arbor hold or flange cradles must be employed for the lift.
- If an inclined ramp is used for unloading, it must be wide enough to contact both flanges completely. The reel should be stopped at the bottom by using the flanges, not the cable surface.
- Never drop reels on the ground.

Storage.

- Reels should be stored on a hard, clean surface, not bare earth.
- Store away from open fires or heat sources.
- Cable ends should be resealed as soon as a cable section has been removed.

Receiving Tests. Cable should be tested according to the insulation type and the methods outlined in Chapter 4 of this handbook.

Installation

Securing and Supporting. Low-voltage cable shall be secured and supported as follows:

- Approved staples, straps, hangers, or similar fittings, as shown in Figure 6–21.
- When mounted on studs, joists, and rafters, cable shall be installed as shown in Figure 6–22.
- When passed through bored holes, the edge of the hole shall be no less than 1¼ inches from the edge of the stud, joist, or rafter.
- In the event that 1¼ inches cannot be maintained, as mentioned previously and shown in Figure 6–3, a steel spacer or sleeve must be used to protect the cable.
- Cable installed in accessible attics shall be secured as shown in Figure 6–23.

High-voltage cable should be installed as described previously for Type MV cable.

Termination and Splicing. Termination of these types is standard to low-voltage applications. See Chapter 3 of this handbook.

NOTE:

> The following cable types are signaling, communications, information, and other types of low-power applications. Because of their unique nature, as compared to power cable, they are included here in their own section. Table 6–6 summarizes the types of cables and their applications.

REMOTE CONTROL, SIGNALING, AND POWER-LIMITED CIRCUIT CABLE (CL*x*)

Description

General. These cable types are used in applications that, by nature of their usage and/or electrical power limitations, differentiate them from standard power circuits. In Article 725, the *NEC*® defines the three classes based on voltage and va output: Classes CL1, CL2, and CL3.

Insulation types may be any material that is listed and approved for this particular class. Conductors of size AWG #16 and AWG #18 may be used but must be

Table 6–6 Communications and signaling cable types.

NEC® Article/Type		Description	Plenum	Riser	Commercial	Residential
725	CL1	Class 1 cables	XX	XX	CL1	XX
	CL2	Class 2 cables	CL2P	CL2R	CL2	CL2X*
	CL3	Class 3 cables	CL3P	CL3R	CL3	CL3X*
	PLTC	A stand-alone class. This is a power limited tray cable—a CL3-type cable that can be used outdoors. Is sunlight and moisture resistant and must pass the vertical tray flame test.	(none)	(none)	PLTC	(none)
760	(N) FPL	(Non)Power-limited, fire-protective signalling circuit cable	FPLP	FPLR	FPL	(none)
770	OFC	Fiber cable also containing metallic conductors	OFCP	OFCR	OFCG, OFC	(none)
	OFN	Fiber cable only containing optical fibers	OFNP	OFNR	OFNG, OFN	(none)
800	CM	Communications	CMP	CMR	CMG, CM	CMX*
	MP	Multipurpose cables	MPP	MPR	MPG, MP	(none)

*Cable diameter must be less than 0.250″.

of Type FFH-2, KF-2, KFF-2, PAF, PAFF, PF, PFF, PGF, PGFF, PTF, PTFF, RFH-2, RFHH-2, RFHH-3, SF-2, SFF-2, TF, TFF, TFFN, TFN, ZF, or ZFF. (See Appendix A for a description of these *NEC*® types.)

Types. See Table 6–6.

Applications

Remote control, signaling, and power-limited applications. These type of circuits are inherently low power type of circuits. See *NEC*® Article 725.

Sizes

From #18 AWG to #10 AWG (Note: There is no practical limit to size. Generally size is determined by current requirements. (See the "Ampacities" section following.)

Temperature Ranges

As determined by insulation type.

Voltage Ranges

Class 1
Power limited circuits—30 volts
Remote control and signaling—600 volts

Classes 2 and 3
See Table 6–7

Ampacities

Determined using standard *NEC*® methods except when fixture wire is used as shown in the following:

Size (AWG)	Ampacity
18	6
16	8
14	17
12	23
10	28

See Table 6–7 for va capacities of Class 2 and Class 3 circuits.

Table 6–7 Class 2 and Class 3 power sources. ***(Reprinted with permission from NFPA 70–1999)***

a. Class 2 and Class 3 Alternating Current Power Source Limitations

Power Source		Inherently Limited Power Source (Overcurrent Protection Not Required)				Not Inherently Limited Power Source (Overcurrent Protection Required)			
		Class 2			Class 3	Class 2		Class 3	
Source Voltage V_{max} (volts) (Note 1)		0 through 20†	Over 30 and through 150	Over 30 and through 150	Over 30 and through 100	0 through 20†	Over 20 and through 30†	Over 30 and through 100	Over 100 and through 150
Power Limitations $(VA)_{max}$ (va) (Note 1)		—	—	—	—	250 (Note 3)	250	250	N/A
Current Limitations I_{max} (amps) (Note 1)		8.0	0.005	0.005	$150/V_{max}$	$1000/V_{max}$	$1000/V_{max}$	$1000/V_{max}$	1.0
Maximum Overcurrent Protection (amps)		—	—	—	—	5.0	$100/V_{max}$	$100/V_{max}$	1.0
Power Source Maximum Nameplate Rating	VA (va)	$5.0 \times V_{max}$	100 ×	$0.005 \times V_{max}$	100	$5.0 \times V_{max}$	100	100	100
	Current (Amps)	5.0	$100/V_{max}$	0.005	$100/V_{max}$	5.0	$100/V_{max}$	$100/V_{max}$	$100/V_{max}$

†Voltage ranges shown are for sinusoidal AC in indoor locations or where wet contact is not likely to occur. For nonsinusoidal or wet contact conditions, see Note 2.

Table 6–7 Class 2 and Class 3 power sources. *(Reprinted with permission from NFPA 70–1999)* *continued*

b. Class 2 and Class 3 Direct Current Power Source Limitations

Power Source	Inherently Limited Power Source (Overcurrent Protection Not Required)					Not Inherently Limited Power Source (Overcurrent Protection Required)			
	Class 2			Class 3		Class 2		Class 3	
Source Voltage V_{max} (volts) (Note 1)	0 through 20††	Over 20 and through 30††	Over 30 and through 60††	Over 60 and through 150	Over 60 and through 100	0 through 20††	Over 20 and through 60††	Over 60 and through 100	Over 100 and through 150
Power Limitations $(VA)_{max}$ (va) (Note 1)	—	—	—	—	—	250 (Note 3)	250	250	NA
Current Limitations I_{max} (amps) (Note 1)	8.0	8.0	$150/V_{max}$	0.005	$150/V_{max}$	$1000/V_{max}$	$1000/V_{max}$	$1000/V_{max}$	1.0
Maximum Overcurrent Protection (amps)	—	—	—	—	—	5.0	$100/V_{max}$	$100/V_{max}$	1.0
Power Source Maximum Nameplate Rating — VA (va)	$5.0 \times V_{max}$	100	100	$0.005 \times V_{max}$	100	$5.0 \times V_{max}$	100	100	100
Power Source Maximum Nameplate Rating — Current (Amps)	5.0	$100/V_{max}$	$100/V_{max}$	0.005	$100/V_{max}$	5.0	$100/V_{max}$	$100/V_{max}$	$100/V_{max}$

††Voltage ranges shown are for continuous DC in indoor locations or where wet contact is not likely to occur. For interrupted DC or wet contact conditions, see Note 4.

Notes for Tables 6–7a and 6–7b

Note 1. V_{max}, I_{max}, and VA_{max} are determined with the current limiting impedance in the circuit (not bypassed) as follows:

V_{max}: Maximum output voltage regardless of load with rated input applied.

I_{max}: Maximum output current under any noncapacitive load, including short circuit, and with overcurrent protection bypassed if used. Where a transformer limits the output current, I_{max} limits apply after 1 minute of operation. Where a current-limiting impedance, listed for the purpose, or as part of a listed product, is used in combination with a nonpower-limited transformer or a stored energy source (e.g., storage battery) to limit the output current, I_{max} limits apply after 5 seconds.

VA_{max}: Maximum va output after 1 minute of operation regardless of load and overcurrent protection bypassed if used.

Note 2: For nonsinusoidal AC, V_{max} shall not be greater than 42.4 volts peak. Where wet contact (immersion not included) is likely to occur, Class 3 wiring methods shall be used or V_{max} shall be not greater than 15 volts for sinusoidal AC, 21.2 volts peak for nonsinusoidal AC.

Note 3. If the power source is a transformer, VA_{max} is 350 or less when V_{max} is 15 or less.

Note 4. For DC interrupted at a rate of 10 to 200 Hz, V_{max} shall not be greater than 24.8 volts peak. Where wet contact (immersion not included) is likely to occur, Class 3 wiring methods shall be used or V_{max} shall not be greater than 30 volts for continuous DC, 12.4 volts peak for DC that is interrupted at a rate of 10 to 200 Hz.

Handling and Care

Handling and care of CL cables should be determined by the type of insulation being employed. Refer to the appropriate insulation section.

Installation

These classes may be installed generally in any way as determined by the insulation class. Note from Table 6–6 that plenum and riser rated classes (suffixes P and R respectively) are designed to install in such locations. They must, therefore, be able to withstand the conditions therein and must be rated for such installation.

Termination and Splicing

Termination depends on the application. Generally the techniques used in Figure 6–24 will apply.

POWER LIMITED TRAY CABLE (PLTC)

PLTC is a CL3 type cable designed especially for use in cable trays. See the previous section and Table 6–6.

NONPOWER LIMITED AND POWER-LIMITED FIRE PROTECTIVE SIGNALING CIRCUIT CABLE (NFPL AND FPL)

Description

General. These cable types are used in fire protection alarm signaling circuits. They may be used as single- or multiple-conductor applications.

NFPL cables must be copper. NFPL insulation types may be any material that is listed and approved for this particular class. Conductors of size AWG #16 and AWG #18 may be used but must be of type KF-2, KFF-2, PAFF, PPTFF, PF, PFF, PGF, PGFF, RFH-2, RFHH-2, RFHH-3, SF-2, SFF-2, TF, TFF, TFN, TFFN, ZF, or ZFF. (See Appendix A for a description of these *NEC*® types.)

FPL cables are required to be supplied by Class 3 power sources and therefore have essentially the same type of requirements as do the Class 3 cables previously listed.

Types. Table 6–6 defines the general types.

Applications

NFPL: Nonpower limited fire alarm signaling circuits.
FPL: Power-limited fire alarm signaling circuits.
See *NEC*® Article 760.

Sizes

NFPL: From #18 AWG to #10 AWG. (Note: There is no practical limit to the size. Generally size is determined by current requirements. See the "Ampacities" section following.)
FPL: Sizes as small as #26 are allowed as defined by *NEC*® Article 760.

Temperature Ranges

As determined by insulation type.

Voltage Ranges

NFPL	FPL
600 volts max	Same as Class 3 (Table 6–7)

Ampacities

Determined using standard *NEC*® methods except when fixture wire is used as shown in the following:

Size (AWG)	Ampacity
18	6
16	8

See Table 6–7 for va capacities of Class 3 circuits.

Handling and Care

Handling and care of NPFL and PFL cables should be determined by the type of insulation being employed. Refer to the appropriate insulation section.

Installation

These classes may be installed generally in any way as determined by the insulation class. Note from Table 6–6 that plenum and riser rated classes (suffixes P and R respectively) are designed to install in such locations. They must, therefore, be able to withstand the conditions therein and must be rated for such installation.

Termination and Splicing

Termination depends on the application. Generally the techniques used in Figure 6–24 will apply.

OPTICAL FIBER CABLE (OFC AND OFN)

Description

General. The OFC and OFN types are nonconductive optical fiber type cables used for communications, signaling, data transfer, and other such applications. Electrical insulation materials are generally unimportant except in those situations in which optical fiber is used close to electrical conductors.

Types. Table 6–6 defines the general types. In addition, the *NEC*® Article 770-5 defines the following subcategories:

Nonconductive: Cables with no conductive materials.

Conductive: Cables containing conductive members that are noncurrent carrying (e.g., metallic strength members).

Composite: Cables containing optical fiber as well as current-carrying electrical conductors. Note that this type may also contain noncurrent-carrying members as described previously.

Applications

These cable types should generally be installed only with other signaling and communications type cables, except as defined in the following:

Nonconductive:
Permitted:

1. Same cable tray or raceway with conductors for electric light, power, Class 1, nonpower-limited fire alarm, or medium power network-powered broadband communications circuits of less than 600 volts. (Note: Optical fiber may be installed in proximity to circuits over 600 volts in industrial environments where only qualified personnel service the installation.)
2. Same cable tray, enclosure, or raceway with:
    a. Class 2 and Class 3 remote-control, signaling, and power-limited circuit. (Article 725).
    b. Power-limited fire alarm systems in compliance with Article 760.
    c. Communications circuits (Article 800).
    d. Community antenna TV and radio distribution systems (Article 820).
    e. Low power network-powered broadband communications circuits in compliance with Article 830.

Nonpermitted: Same cabinet, outlet box, panel, or similar enclosure with conductors for electric light, power, Class 1, nonpower-limited fire alarm, or medium power network-powered broadband communications circuits. Unless:

1. All cables are functionally associated.
2. If the optical fiber cables are installed in a factory or field-assembled control center.

Conductive:

Permitted: Same cable tray, enclosure, or raceway with:

a. Class 2 and Class 3 remote-control, signaling, and power-limited circuits.
b. Power-limited fire alarm systems in compliance with Article 760.
c. Communications circuit (Article 800).
d. Community antenna TV and radio distribution systems (Article 820).
e. Low power network-powered broadband communications circuits in compliance with Article 830.

Nonpermitted: Same cable tray or raceway with conductors for electric light, power, Class 1, nonpower-limited fire alarm, or medium power network-powered broadband communications circuits.

Composite:

Permitted:

1. Optical fibers are permitted in a composite cable that carries conductors for electric light, power, Class 1, nonpower-limited fire alarm, or medium power network-powered broadband communications systems operating at 600 volts or less if all conductors are associated with the same purpose.
2. In the same cabinet, cable tray, outlet box, panel, raceway, or other termination enclosure with conductors for electrical light, power, or Class 1 circuits operating at 600 volts or lower. Exception: In industrial establishments, where only qualified persons will service the installation, composite cables may contain conductors in excess of 600 volts.

Sizes

N/A.

Temperature Ranges

As determined by insulation type.

Voltage Ranges

N/A.

Ampacities

N/A.

Handling and Care

Refer to manufacturer's instructions.

Installation

These classes may be installed generally in any way as determined by the insulation class. Note from Table 6–6 that plenum and riser rated classes (suffixes P and R respectively) are designed to install in such locations. They must, therefore, be able to withstand the conditions therein and must be rated for such installation.

Termination and Splicing

See Chapter 3.

NOTE:

There is no practical limit to the size. Generally size is determined by current requirements.

COMMUNICATIONS AND MULTIPURPOSE CABLES (CM*x* and MP*x*)

Description

General. These cable types are used in communications (CM*x*) and/or multiple-purpose applications (MP*x*). Insulation types are generally as defined for the particular application to which the cable is being placed. For example, if used in a Class 2 signaling circuit, the cable would have to have such an insulation.

Types. See Table 6–6.

Applications

Communications, remote control, signaling, and other such power-limited applications. These type of circuits are inherently low power type of circuits. See *NEC®* Article 800.

Sizes

From #18 AWG to #10 AWG

Temperature Ranges

As determined by insulation type.

Voltage Ranges

Generally determined by insulation type and application. However, *NEC®* Article 800 puts a top rating of 600 volts.

Ampacities

Determined by standard *NEC®* methods.

Handling and Care

Handling and care of CL cables should be determined by the type of insulation being employed. Refer to the appropriate insulation section.

Installation

These classes may be installed generally in any way as determined by the insulation class. Note from Table 6–6 that plenum and riser rated classes (suffixes P and R respectively) are designed to install in such locations. They must, therefore, be able to withstand the conditions therein and must be rated for such installation.

Termination and Splicing

Termination depends on the application. Generally the techniques used in Figure 6–24 will apply.

Appendix A

Appendix Table 402–3 for *NEC*®

Trade Name	Type Letter	Insulation	AWG	Thickness of Insulation	Mils	Outer Covering	Maximum Operating Temperature	Application Provisions
				Thickness of Moisture-Resistant Insulation Mils	Thickness of Asbestos Mils			
Asbestos Covered Heat-Resistant Fixture Wire	AF	Impregnated Asbestos or Moisture-Resistant Insulation and Impregnanted Asbestos	18–14 12–10	— 20 — 25	30 10 45 20	None	150°C 302°F	Fixture wiring. Limited to 300 volts and indoor dry locations.
Heat-Resistant Rubber-Covered Fixture Wire—Flexible Standing	FFH–2	Heat-Resistant Rubber Heat-Resistant Latex Rubber	18–16 18–16	. . . . 30 18		Nonmetallic Covering	75° C 167° F	Fixture wiring.
ECTFE Solid or 7-Strand	HF	Ethylene Chloro-Trifluoro-Ethylene	18–14	 15		None	150° F 302° F	Fixture wiring.
ECTFE Flexible Stranding	HFF	Ethylene Chloro-Trifluoro-Ethylene	18–14	 15		None	150° C 302° F	Fixture wiring.
Tape Insulated Fixture Wire—Solid or 7-Strand	KF–1	Aromatic Polyimide Tape	18–10	. . . . 5.5		None	200° C 392° F	Fixture wiring. Limited to 300 volts.
	KF–2	Aromatic Polyimide Tape	18–10	. . . . 8.4		None	200° C 392° F	Fixture wiring.
Tape Insulated Fixture Wire—Flexible Stranding	KFF–1	Aromatic Polyimide Tape	18–10	. . . . 5.5		None	200° C 392° F	Fixture wiring. Limited to 300 volts.
	KFF–2	Aromatic Polyimide Tape	18–10	. . . . 8.4		None	200° C 392° F	Fixture wiring.
Perfluoroalkoxy—Solid or 7-Strand (Nickel or Nickel-Coated Copper)	PAF	Perfluoroalkoxy	18–14	 20		None	250° C 482° F	Fixture wiring (nickel or nickel-coated copper).
Perfluoroalkoxy—Flexible Stranding	PAFF	Perfluoroalkoxy	18–14	 20		None	150° C 302° F	Fixture wiring.

Appendix Table 402–3 for *NEC*®

Trade Name	Type Letter	Insulation	AWG	Thickness of Insulation	Mils	Outer Covering	Maximum Operating Temperature	Application Provisions
Fluorinated Ethylene Propylene Fixture Wire—Solid or 7-Strand	PF	Fluorinated Ethylene Propylene	18–14	....	20	None	200° C 392° F	Fixture wiring. Limited to 300 volts.
Fluorinated Ethylene Propylene Fixture Wire—Flexible Stranding	PFF	Fluorinated Ethylene Propylene	18–14	....	20	None	150° C 302° F	Fixture wiring.
Fluorinated Ethylene Propylene Fixture Wire—Solid or 7-Strand	PGF	Fluorinated Ethylene Propylene	18–14	....	14	Glass braid	200° C 482° F	Fixture wiring.
Fluorinated Ethylene Propylene Fixture Wire—Flexible Stranding	PGFF	Fluorinated Ethylene Propylene	18–14	....	14	Glass braid	150° C 302° F	Fixture wiring (nickel or nickel-coated copper).
Extruded Polytetra-Fluoroethylene Solid or 7-Strand (Nickel or Nickel-Coated Copper)	PTF	Extruded Polytetra-Fluoroethylene	18–14	....	20	None	200° C 482° F	Fixture wiring (silver or nickel-coated copper).
Extruded Polytetra-Fluoroethylene Flexible Stranding 26-36 AWG Silver or Nickel-Coated Copper)	PTFF	Extruded Polytetra-Fluoroethylene	18–14	....	20	None	150° C 302° F	Fixture wiring. Limited to 300 volts.
Heat-Resistant Rubber-Covered Fixture Wire—Solid or 7-Strand	RFH–1	Heat-Resistant Rubber	18	....	15	Nonmetallic	75° C 167° F	Fixture wiring.
	RFH–2	Heat-Resistant Rubber	18–16	....	30	Nonmetallic	75° C 167° F	Fixture wiring. Multiconductor cable.
		Heat-Resistant Latex Rubber	18–16	....	18			
Heat-Resistant Cross-Linked Synthetic Polymer-Insulated Fixture Wire—Solid or Stranded	RFHH–2*	Cross-Linked Synthetic Polymer	18–16	....	30	None or Nonmetallic	90° C 194° F	Fixture wiring.
	RFHH–3*		18–16	....	45			
Silicone Insulated Fixture Wire—Solid or 7-Strand	SF–1	Silicone Rubber	18	....	15	Nonmetallic	200° C 392° F	Fixture wiring. Limited to 300 volts.
	SF–2	Silicone Rubber	18–14	....	30	Nonmetallic	200° C 392° F	Fixture wiring.

*Insulations and outer coverings that meet the requirements of flame-retardant, limited smoke and are so listed shall be permitted to be designated limited smoke with the suffix LS after the Code type designation.

Appendix Table 402–3 for *NEC*®

Trade Name	Type Letter	Insulation	AWG	Thickness of Insulation Mils	Outer Covering	Maximum Operating Temperature	Application Provisions
Silicone Insulated Fixture Wire—Flexible Stranding	SFF–1	Silicone Rubber	18	 15	Nonmetallic Covering	150° C 302° F	Fixture wiring. Limited to 300 volts.
	SFF–2	Silicone Rubber	18–14	 30	Nonmetallic	150° C 302° F	Fixture wiring.
Thermoplastic Covered Fixture Wire—Solid or 7-Strand	TF*	Thermoplastic	18–16	 30	None	60° C 140° F	Fixture wiring.
Thermoplastic Covered Fixture Wire—Flexible Stranding	TFF*	Thermoplastic	18–16	 30	None	60° C 140° F	Fixture wiring.
Heat-Resistant Thermoplastic-Covered Fixture Wire—Solid or 7-Strand	TFN*	Thermoplastic	18–16	 15	Nylon-Jacketed or equivalent	90° C 194° F	Fixture wiring.
Heat-Resistant Thermoplastic-Covered Fixture Wire—Flexible Stranded	TFFN*	Thermoplastic	18–16	 15	Nylon-Jacketed or equivalent	90° C 194° F	Fixture wiring.
Cross-Linked Polyolefin Insulated Fixture Wire—Solid or 7-Strand	XF*	Cross-Linked Polyolefin	18–14 12–10	. . . . 30 45	None	150° C 302° F	Fixture wiring. Limited to 300 volts.
Cross-Linked Polyolefin Insulated Fixture Wire—Flexible Stranded	XFF*	Cross-Linked Polyolefin	18–14 12–10	. . . . 30 45	None	150° C 302° F	Fixture wiring. Limited to 300 volts.
Modified ETFE Solid or 7-Strand	ZF	Modified Ethylene Tetrafluoro-Ethylene	18–14	 15	None	150° C 302° F	Fixture wiring.
Flexible Stranding	ZFF	Modified Ethylene Tetrafluoro-Ethylene	18–14	 15	None	150° C 302° F	Fixture wiring.
High Temp. Modified ETFE—Solid or 7-Strand	ZHF	Modified Ethylene Tetrafluoro-Ethylene	18–14	 15	None	200° C 392° F	Fixture wiring.

*Insulations and outer coverings that meet the requirements of flame-retardant, limited smoke and are so listed shall be permitted to be designated limited smoke with the suffix LS after the Code type designation.

Index

A

B

C

W

X

Z